Lecture Notes in Earth Sciences

Editors:

J. Reitner, Göttingen
M. H. Trauth, Potsdam
K. Stüwe, Graz
D. Yuen, USA

Founding Editors:

G. M. Friedman, Brooklyn and Troy
A. Seilacher, Tübingen and Yale

Chongbin Zhao · Bruce E. Hobbs · Alison Ord

Fundamentals of Computational Geoscience

Numerical Methods and Algorithms

Dr. Chongbin Zhao
Computational Geosciences
Research Centre
Central South University
Changsha, China
Chongbin.zhao@iinet.net.au

Dr. Bruce E. Hobbs
School of Earth and Geographical Sciences
The University of Western Australia
Perth, Australia
Bruce.Hobbs@csiro.au

Dr. Alison Ord
School of Earth and Geographical Sciences
The University of Western Australia
Perth, Australia
and CSIRO Division of
Exploration and Mining
Perth, Australia
alison.ord@csiro.au

ISBN 978-3-540-89742-2   e-ISBN 978-3-540-89743-9

DOI 10.1007/978-3-540-89743-9

Lecturer Notes in Earth Sciences ISSN 0930-0317

Library of Congress Control Number: 2008942068

© Springer-Verlag Berlin Heidelberg 2009

This work is subject to copyright. All rights are reserved, whether the whole or part of the material is concerned, specifically the rights of translation, reprinting, reuse of illustrations, recitation, broadcasting, reproduction on microfilm or in any other way, and storage in data banks. Duplication of this publication or parts thereof is permitted only under the provisions of the German Copyright Law of September 9, 1965, in its current version, and permission for use must always be obtained from Springer. Violations are liable to prosecution under the German Copyright Law.

The use of general descriptive names, registered names, trademarks, etc. in this publication does not imply, even in the absence of a specific statement, that such names are exempt from the relevant protective laws and regulations and therefore free for general use.

Cover design: deblik, Berlin

Printed on acid-free paper

9 8 7 6 5 4 3 2 1

springer.com

Acknowledgements

Dr Chongbin Zhao expresses his sincere thanks to his wife, Ms Peiying Xu, for her persistent support and encouragement, without which it would be impossible to write this monograph. We are very grateful to the Central South University for financial support during writing this monograph. The partial work of this monograph is also financially supported by the Natural Science Foundation of China (Grant Nos: 10872219 and 10672190). We express our thanks to the anonymous referees for their valuable reviews of this monograph.

Preamble

In recent years, numerical methods and computational simulations provide a new way to deal with many geoscience problems, for which the traditionally-used theoretical and experimental methods may not be valid as a result of the large time and length scales of the problems themselves. This enables many hitherto unsolvable geoscience problems to be solved using numerical methods and computational simulations. In particular, through wide application of computational science to geoscience problems, a new discipline, namely computational geoscience, has been established. However, because of the extremely large length and time scales, the numerical simulation of a real geological world also provides many challenging problems for researchers involved in the field of computational science. For this reason, multidisciplinary knowledge and expertise from mathematicians, physicists, chemists, computational scientists and geoscientists are required in the process of establishing the research methodology of computational geoscience.

Since computational geoscience is an amalgamation of geoscience and computational science, theoretical analysis and computational simulation are two of its core members. On the theoretical analysis front, we need: (1) to measure and gather data and information through traditional geoscience observations and measurements such as those widely used in geology, geophysics, geochemistry and many other scientific and engineering fields; (2) to conduct research to find the key factors and processes that control the geoscience problem under consideration; (3) to establish the theoretical foundations of the geoscience problem through formulating a set of partial differential equations on the basis of fundamental scientific principles; (4) to investigate the solution characteristics of these partial differential equations using rigorous mathematical treatments. On the computational simulation front, we need: (1) to develop advanced numerical methods, procedures and algorithms for simulating multi-scale and multi-process aspects of the geoscience problem on the basis of contemporary computational science knowledge and expertise; (2) to verify computational codes established on the basis of these advanced numerical methods, procedures and algorithms through comparing numerical solutions with benchmark solutions; (3) to produce and validate numerical solutions of real geoscience problems.

Owing to the broad nature of geoscience problems, computational geoscience is at a developing stage. Nevertheless, under the stimulus of ever-increasing demand for natural mineral resources, computational geoscience has achieved much in the past decade, driven from the need to understand controlling mechanisms behind ore body formation and mineralization in hydrothermal systems within the upper crust of the Earth. In order to disseminate widely the existing knowledge of computational geoscience, to promote extensively and fastly further development of the computational geoscience, and to facilitate efficiently the broad applications of computational geoscience, it is high time to publish a monograph to report the current knowledge in a systematic manner. This monograph aims to provide state-of-the-art numerical methods, procedures and algorithms in the field of computational geoscience, based on the authors' own work during the last decade. For this purpose, although some theoretical results are provided to verify numerical ones, the main focus of this monograph is on computational simulation aspects of this newly-developed computational geoscience discipline. The advanced numerical methods, procedures and algorithms contained in this monograph are also applicable to a wide range of problems of other length-scales such as engineering length-scales. To broaden the readership of this monograph, common mathematical notations are used to describe the theoretical aspects of geoscience problems. This enables this monograph to be used either as a useful textbook for postgraduate students or as an indispensable reference book for computational geoscientists, mathematicians, engineers and geoscientists. In addition, each chapter is written independently of the remainder of the monograph so that readers may read the chapter of interest separately.

In this monograph we use the finite element method, the finite difference method and the particle simulation method as basic numerical methods for dealing with geoscience problems. Not only have these three methods been well developed in the field of computational science, but also they have been successfully applied to a wide range of small-scale scientific and engineering problems. Based on these three methods, we have developed advanced numerical procedures and algorithms to tackle the large-scale aspects of geoscience problems. The specific geoscience problem under consideration is the ore body formation and mineralization problem in hydrothermal systems within the upper crust of the Earth. Towards this end, we present the advanced procedures and algorithms in this monograph as follows: (1) Due to the important role that convective pore-fluid flow plays in the controlling processes of ore body formation and mineralization, a progressive asymptotic approach procedure is proposed to solve steady-state convective pore-fluid flow problems within the upper crust of the Earth. (2) To consider both the thermoelastic effect and the double diffusion effect, a consistent point-searching interpolation algorithm is proposed to develop a general interface between two commercial computer codes, Fluid Dynamics Analysis Package (FIDAP) and Fast Lagrangian Analysis of Continua (FLAC). This general interface allows a combination use of the two commercial codes for solving coupled problems between medium deformation, pore-fluid flow, heat transfer and reactive mass transport processes that can occur simultaneously in hydrothermal systems. (3) To simulate mineral dissolution/precipitation and metamorphic processes, a term splitting algorithm is developed for dealing with fluid-rock interaction problems in fluid-saturated hydrothermal/sedimentary basins of subcritical Zhao numbers, in which the chemical dissolution fronts are stable during their propagation. Note that the Zhao number is a dimensionless number that can be used to represent the geometrical, hydrodynamic, thermodynamic and chemical kinetic characteristics of a reactive transport system in a comprehensive manner. The condition, under which a chemical dissolution front in the fluid-saturated porous medium becomes unstable, can be expressed by the critical value of this dimensionless number. (4) For a geochemical system of critical and supercritical Zhao numbers, a segregated algorithm is proposed for solving chemical-dissolution front instability problems in fluid-saturated porous rocks. Thus, the morphological evolution of chemical dissolution fronts in fluid-saturated porous media can be appropriately simulated. (5) To investigate the effects of non-equilibrium redox chemical reactions on the mineralization patterns in hydrothermal systems, a decoupling procedure is proposed for simulating fluids mixing, heat transfer and non-equilibrium redox chemical reactions in fluid-saturated porous rocks. (6) When thermal and chemical effects of intruded magma are taken into account, an equivalent source algorithm is presented for simulating thermal and chemical effects of intruded magma solidification problems. This algorithm enables the moving boundary problem associated with magma solidification to be effectively and efficiently solved using the fixed finite element meshes. (7) To simulate spontaneous crack generation in brittle rocks within the upper crust of the Earth, the particle simulation method is extended to solve spontaneous crack generation problems associated with faulting and folding in large length-scale geological systems. The resulting cracks may be connected to form flow channels, which can control ore body formation and mineralization patterns within the upper crust of the Earth.

August 12, 2008

Chongbin Zhao
Bruce E. Hobbs
Alison Ord

Contents

1 **Introduction** 1
   1.1 Characteristics of Computational Geoscience 2
   1.2 Basic Steps Associated with the Research Methodology
        of Computational Geoscience 3
        1.2.1 The Conceptual Model of a Geoscience Problem 3
        1.2.2 The Mathematical Model of a Geoscience Problem 3
        1.2.3 The Numerical Simulation Model of a Geoscience Problem .. 4
        1.2.4 Graphical Display of the Numerical Simulation Results 5
   1.3 The Contextual Arrangements of this Monograph 5

2 **A Progressive Asymptotic Approach Procedure for Simulating
Steady-State Natural Convective Problems in Fluid-Saturated
Porous Media** 7
   2.1 Governing Equations of the Problem 9
   2.2 Finite Element Formulation of the Problem 11
   2.3 The Progressive Asymptotic Approach Procedure
        for Solving Steady-State Natural Convection Problems
        in Fluid-Saturated Porous Media 14
   2.4 Derivation of Analytical Solution to a Benchmark Problem 16
   2.5 Verification of the Proposed Progressive Asymptotic Approach
        Procedure Associated with Finite Element Analysis 19
   2.6 Application of the Progressive Asymptotic Approach Procedure
        Associated with Finite Element Analysis 22
        2.6.1 Two-Dimensional Convective Pore-Fluid Flow Problems 22
        2.6.2 Three-Dimensional Convective Pore-Fluid Flow Problems ... 28

3 **A Consistent Point-Searching Interpolation Algorithm for
Simulating Coupled Problems between Deformation, Pore-Fluid
Flow, Heat Transfer and Mass Transport Processes in Hydrothemal
Systems** 37
   3.1 Statement of the Coupled Problem and Solution Method 38
   3.2 Mathematical Formulation of the Consistent Point-Searching
        Interpolation Algorithm in Unstructured Meshes 42

xi

		3.2.1 Point Searching Step 43
		3.2.2 Inverse Mapping Step 45
		3.2.3 Consistent Interpolation Step 50
	3.3	Verification of the Proposed Consistent Point-Searching Interpolation Algorithm 51
	3.4	Application Examples of the Proposed Consistent Point-Searching Interpolation Algorithm 56
		3.4.1 Numerical Modelling of Coupled Problems Involving Deformation, Pore-Fluid Flow and Heat Transfer in Fluid-Saturated Porous Media 56
		3.4.2 Numerical Modelling of Coupled Problems Involving Deformation, Pore-Fluid Flow, Heat Transfer and Mass Transport in Fluid-Saturated Porous Media 59

4 A Term Splitting Algorithm for Simulating Fluid-Rock Interaction Problems in Fluid-Saturated Hydrothermal Systems of Subcritical Zhao Numbers 73

	4.1	Key Issues Associated with the Numerical Modelling of Fluid-Rock Interaction Problems 76
	4.2	Development of the Term Splitting Algorithm 77
	4.3	Application Examples of the Term Splitting Algorithm 82

5 A Segregated Algorithm for Simulating Chemical Dissolution Front Instabilities in Fluid-Saturated Porous Rocks 95

	5.1	Mathematical Background of Chemical Dissolution Front Instability Problems in Fluid-Saturated Porous Rocks 96
		5.1.1 A General Case of Reactive Multi-Chemical-Species Transport with Consideration of Porosity/Permeability Feedback 96
		5.1.2 A Particular Case of Reactive Single-Chemical-Species Transport with Consideration of Porosity/Permeability Feedack 99
	5.2	Proposed Segregated Algorithm for Simulating the Morphological Evolution of a Chemical Dissolution Front 109
		5.2.1 Formulation of the Segregated Algorithm for Simulating the Evolution of Chemical Dissolution Fronts 109
		5.2.2 Verification of the Segregated Algorithm for Simulating the Evolution of Chemical Dissolution Fronts 111
	5.3	Application of the Segregated Algorithm for Simulating the Morphological Evolution of Chemical Dissolution Fronts 115

6 A Decoupling Procedure for Simulating Fluid Mixing, Heat Transfer and Non-Equilibrium Redox Chemical Reactions in Fluid-Saturated Porous Rocks ... 121

- 6.1 Statement of Coupled Problems between Fluids Mixing, Heat Transfer and Redox Chemical Reactions ... 123
- 6.2 A Decoupling Procedure for Removing the Coupling between Reactive Transport Equations of Redox Chemical Reactions ... 126
- 6.3 Verification of the Decoupling Procedure ... 128
- 6.4 Applications of the Proposed Decoupling Procedure to Predict Mineral Precipitation Patterns in a Focusing and Mixing System Involving Two Reactive Fluids ... 134
  - 6.4.1 Key Factors Controlling Mineral Precipitation Patterns in a Focusing and Mixing System Involving Two Reactive Fluids ... 136
  - 6.4.2 Theoretical Analysis of Mineral Precipitation Patterns in a Focusing and Mixing System Involving Two Reactive Fluids ... 138
  - 6.4.3 Chemical Reaction Patterns due to Mixing and Focusing of Two Reactive Fluids in Permeable Fault Zones ... 140
  - 6.4.4 Numerical Illustration of Three Types of Chemical Reaction Patterns Associated with Permeable Fault Zones ... 145

7 An Equivalent Source Algorithm for Simulating Thermal and Chemical Effects of Intruded Magma Solidification Problems ... 153

- 7.1 An Equivalent Source Algorithm for Simulating Thermal and Chemical Effects of Intruded Magma Solidification Problems ... 155
- 7.2 Implementation of the Equivalent Source Algorithm in the Finite Element Analysis with Fixed Meshes ... 160
- 7.3 Verification and Application of the Equivalent Source Algorithm ... 163

8 The Particle Simulation Method for Dealing with Spontaneous Crack Generation Problems in Large-Scale Geological Systems ... 175

- 8.1 Basic Formulations of the Particle Simulation Method ... 179
- 8.2 Some Numerical Simulation Issues Associated with the Particle Simulation Method ... 184
  - 8.2.1 Numerical Simulation Issue Caused by the Difference between an Element and a Particle ... 184
  - 8.2.2 Numerical Simulation Issue Arising from Using the Explicit Dynamic Relaxation Method to Solve a Quasi-Static Problem ... 186
  - 8.2.3 Numerical Simulation Issue Stemming from the Loading Procedure Used in the Particle Simulation Method ... 189
- 8.3 An Upscale Theory of Particle Simulation for Two-Dimensional Quasi-Static Problems ... 194

8.4 Test and Application Examples of the Particle Simulation Method ... 199
    8.4.1 Comparison of the Proposed Loading Procedure with the Conventional Loading Procedure 201
    8.4.2 The Similarity Test of Two Particle Samples of Different Length-Scales 204
    8.4.3 Particle Simulation of the Folding Process Using Two Similar Particle Models of Different Length-Scales 210
    8.4.4 Particle Simulation of the Faulting Process Using the Proposed Particle Method 216

Summary Statements 221

References 227

Index 239

Nomenclature

The following symbols are commonly used with the attached definitions, unless otherwise specified in the monograph.

A	area of a finite element
C	species concentration
$\mathbf{C}$	species concentration vector
C_1	non-zero constant
C_2	arbitrary constant
c_p	specific heat of pore-fluid
D	mass diffusivity
g	acceleration due to gravity
H	reference length
K	medium permeability
K_h	reference medium permeability in the horizontal direction
L	length of a problem domain
Le	Lewis number
P	pressure
$\mathbf{P}$	pressure vector
P_0	hydrostatic pressure
q_c	mass flux on the boundary of a finite element
q_T	heat flux on the boundary of a finite element
Ra	Rayleigh number
$Ra_{critical}$	critical Rayleigh number
S	boundary length of a finite element
T	temperature
$\mathbf{T}$	temperature vector
t	temporal variable
u	Darcy velocity in the x direction
$\mathbf{U}$	Darcy velocity vector
v	Darcy velocity in the y direction
w	Darcy velocity in the z direction
x, y, z	spatial coordinates
Zh	Zhao number

$Zh_{critical}$	critical Zhao number
λ	thermal conductivity
λ_{e0}	reference thermal conductivity in the horizontal direction
ϕ	porosity
ψ	stream function
$\boldsymbol{\psi}$	shape function vector for the pressure of a finite element
ρ_0	reference density of pore-fluid
μ	dynamic viscosity of pore-fluid
β	thermal volume expansion coefficient of pore-fluid
σ	stress on the boundary of a finite element
$\boldsymbol{\varphi}$	shape function vector for the temperature, species concentration and Darcy velocity of a finite element
η	permeability ratio of the underlying medium to its overlying folded layer
ζ	thermal conductivity ratio of the underlying medium to its overlying folded layer
ε	penalty parameter associated with the penalty finite element approach

Subscripts

f	pertaining to pore-fluid
0	pertaining to reference quantities

Superscripts

e	pertaining to equivalent quantities of a porous medium
$\mathbf{e}$	pertaining to quantities in a finite element level
$*$	pertaining to dimensionless quantities
s	pertaining to solid matrix

Chapter 1
Introduction

Geoscience is a fundamental natural science discipline dealing with the origin, evolutionary history and behaviour of the planet Earth. As a result of its complicated and complex nature, the Earth system not only provides the necessary materials and environment for mankind to live, but also brings many types of natural disasters, such as earthquakes, volcanic eruptions, tsunamis, floods and tornadoes, to mention just a few. With the ever-increasing demand for improving our living standards, it has been recognized that the existing natural resources will be exhausted in the near future and that our living environments are, in fact, deteriorating. To maintain the sustainable development of our living standards and the further improvement of our living environments, an inevitable and challenging task that geoscientists are now confronting is how accurately to predict not only the occurrences of these natural disasters, but also the locations of large concealed natural resources in the deep Earth. For this reason, geoscientists must study the processes, rules and laws, by which the Earth system operates, instead of simply describing and observing geoscience phenomena. Specifically, geoscientists need to make greater efforts in the following aspects relevant to solving contemporary geoscience problems: (1) the complicated and complex interactions between multi-scales and multi-processes occurring in the solid Earth; (2) gather, accumulate and analyze the large amount of information and data that are essential to understand each of the controlling processes within the interior of the Earth using modern observation equipment, measurement tools, experimental instruments and information processing techniques; (3) the intimate interplay between the solid Earth, biosphere, hydrosphere and atmosphere. It is this intimate interplay that controls the global behaviour of the Earth system. As a result, geoscientists must adopt scientific and predictive methods relevant to conduction of contemporary geoscience research, instead of simply using the traditional descriptive methods.

Computational science is a modern technological science discipline dealing with the development and application of numerical methods, procedures, algorithms and other numerical techniques for delivering numerical solutions for complicated and complex scientific and engineering problems. With the rapid advances and developments of modern computer technology, applications of computational science have penetrated almost all engineering fields: from the topological optimization of a tooth brush to that of a giant aircraft; from the collapse simulation of a concrete beam to that of a huge double curvature arch dam; from the optimal layout design of a pipeline to that of a large-scale underground tunnel, to name just a few. Since computational science is a comprehensive discipline bringing geology, geophysics, geochemistry, mathematics, physics, chemistry, biology and numerical techniques together, it can be used effectively and efficiently to simulate the processes involved in complicated scientific and engineering problems in a scientific and predictive manner. In this sense, computational science is a natural supplier to meet the demands of geoscientists in solving contemporary geoscience problems. It is this demand and supply relationship that has created a brand new discipline, computational geoscience, in the past decade (Zhao et al. 2008a).

1.1 Characteristics of Computational Geoscience

Computational geoscience is a newly-developed discipline, which has been established through applying the well-developed computational science discipline to solve geoscience problems occurring in nature. This means that the computational geoscience discipline is of multi-disciplinary nature crossing many fields of science. The ultimate aim of computational geoscience is to deal with the origin, evolution and behaviour of the Earth system in a predictive, scientific manner. Under the stimulus of an ever-increasing demand for natural mineral resources, computational geoscience has achieved, in the past decade, considerable development driven from the need to understand the controlling mechanisms behind ore body formation and mineralization in hydrothermal and igneous systems within the upper crust of the Earth (Garven and Freeze 1984, Raffensperger and Garven 1995, Doin et al. 1997, Jiang et al. 1997, Zhao et al. 1997a, 1998a, Oliver et al. 1999, 2001, Zhao et al. 1999a, 2000a, Hobbs et al. 2000, Gow et al. 2002, Ord et al. 2002, Schaubs and Zhao 2002, Sorjonen-Ward et al. 2002, Zhao et al. 2002a, 2003a, McLellan et al. 2003, Ord and Sorjonen-Ward 2003, Liu et al. 2005, Sheldon and Ord 2005, Zhao et al. 2005a, 2006a, b, 2007a, 2008a, Zhang et al. 2007, Murphy et al. 2008). As a result, a fundamental and theoretical framework for the computational geoscience discipline has been established. This enables many hitherto unsolvable geoscience problems to be solved, both theoretically and practically, using the newly-developed research methodology associated with computational geoscience. For instance, some typical examples of applying the newly-developed research methodology to deal with geoscience problems are as follows: (1) the convective flow of pore-fluid within the upper crust of the Earth (Phillips 1991, Nield and Bejan 1992, Zhao et al. 1997a, 1998b, 1999b, 2000b, 2001b, Lin et al. 2003), (2) ore body formation and mineralization within hydrothermal systems (Zhao et al. 1998a, 1999c, 2000c, Gow et al. 2002, Ord et al. 2002, Zhao et al. 2002b, 2003b, 2006c), (3) pore-fluid flow focusing within permeable faults (Obdam and Veling 1987, Zimmerman 1996, Zhao et al. 1999d, 2006d, e, 2008b, c), (4) fluid-rock/chemical interaction associated with ore body formation processes (Steefel and Lasaga 1994, Zhao et al. 2001c, 2008d, e, f) and (5) convective flow of pore-fluid within three-dimensional permeable faults (Zhao et al. 2003c, d, 2004, 2005b, Yang 2006).

1.2 Basic Steps Associated with the Research Methodology of Computational Geoscience

Generally speaking, the research methodology of computational geoscience is a comprehensive research methodology, which is formed by combining field observation, theoretical analysis, numerical simulation and field validation. The primary aim of using this research methodology is to investigate the dynamic processes and mechanisms involved in an observed geological phenomenon, rather than to describe the observed geological phenomenon itself. The appropriate research methodology of computational geoscience is usually comprised of the following four main steps: (1) the establishment of a conceptual model for a given geoscience problem; (2) the establishment of a mathematical model for the given problem; (3) the construction of a numerical simulation model for the given problem and (4) the graphical display of the numerical results obtained from the numerical simulation.

1.2.1 The Conceptual Model of a Geoscience Problem

Based on extensive data and information obtained from field and laboratory investigations of a geological phenomenon, a conceptual model is established, which reflects the geometrical architecture and main processes associated with the phenomenon. This is the key step in the process of using the research methodology of computational geoscience to solve a problem. Due to the multiple processes and multiple scales involved in a typical geoscience problem, only the major controlling dynamic processes and mechanisms associated with the problem need to be considered during the initial establishment of the conceptual model. Since other unimportant, or at least less critical, processes and factors are neglected, the initial conceptual model of the problem is somewhat simplified. This conceptual model is used to represent the main characteristics of the real geoscience problem. As understanding grows more detail may be added if necessary.

The fundamental principle involved in establishing the conceptual model for the problem is that the details of the conceptual model should depend on both the length-scale and time-scale of the problem. The conceptual model cannot be over simplified, since then it cannot be used effectively to reflect the main dynamic processes and mechanisms of the real problem. On the other hand, the conceptual model cannot be over complicated, for then unnecessary problems for both the theoretical analysis and the numerical simulation may arise and it may become difficult to unravel which parts of the description of the problem are important.

1.2.2 The Mathematical Model of a Geoscience Problem

Using three fundamental principles, namely the conservation of mass, the conservation of momentum and the conservation of energy, as well as the related physical and chemical laws (Bear 1972, Bear and Bachmat 1990, Phillips 1991, Nield and Bejan 1992, Zhao et al. 1997a), the conceptual model for the given problem can be translated into a mathematical model, which is usually comprised of a set of partial differential equations. Due to the complex and complicated nature of these equations, it is very difficult, if not impossible, to find analytical solutions. Alternatively, numerical simulation methods need to be used to find approximate solutions for the problem.

To ensure the accuracy and reliability of the numerical simulation solution, it is necessary to investigate the solution characteristics of the partial differential equations through a theoretical analysis. For example, some theoretical methods can be used to investigate the solution singularity and multiple solution characteristics of the partial differential equations, as well as the conditions under which such characteristics can occur. If possible, a benchmark model should be established for a particular kind of geoscience problem. The geometrical nature and boundary conditions of this benchmark model can be further simplified, so that the theoretical solution, known as the benchmark solution, can be obtained. This benchmark solution is valuable and indispensable for the verification of both the numerical algorithm and the computer code, which are used to solve the problem that is generally characterised by a complicated geometrical shape and complex material properties. It must be pointed out that, due to the approximate nature of a numerical method, the theoretical investigation of the solution characteristics associated with the partial differential equations of a problem plays an important role in applying the research methodology of computational geoscience to solve real problems. This is the key step to ensure the accuracy and reliability of the numerical solution obtained from the numerical simulation of the problem.

1.2.3 The Numerical Simulation Model of a Geoscience Problem

From a mathematical point of view, the numerical simulation model of a geoscience problem can be also called the discretized type of mathematical model. Both the finite element method and the finite difference method are commonly-used discretization methods for numerical simulation of geoscience problems (Zienkiewicz 1977, Zhao et al. 1998a, 2006a). The basic idea behind these numerical methods is to translate the partial differential equations used to describe the geoscience problem in a continuum system, into the corresponding algebraic equations in a discretized system, which in turn is usually comprised of a large number of elements. Through solving the resulting algebraic equations of the discretized system, a numerical solution can be obtained for the problem.

Compared with an engineering problem, a geoscience problem commonly has both large length-scale and large time-scale characteristics. The length-scale of a geoscience problem is commonly measured in either tens of kilometers or even hundreds of kilometers, while the time-scale of a typical problem is often measured in several million years or even several tens of million years. In addition, most geoscience problems are coupled across both multiple processes and multiple scales. Due to these significant differences between engineering problems and geoscience problems, commercial computer programs and related algorithms, which are mainly

1.3 The Contextual Arrangements of this Monograph   5

designed for solving engineering problems, cannot be directly used to solve geoscience problems without modification. For this reason, it is necessary either to develop new computer programs for solving geoscience problems or to modify the existing commercial computer programs, originally designed for solving engineering problems, so as to be suitable for these problems.

It is noted that the solution reliability of a geoscience problem is strongly dependent on algorithm convergence, algorithm stability, mesh shape, time-step and other factors. To ensure the accuracy and reliability of the computational simulation result for a problem, the above-mentioned factors need to be carefully considered in the process of establishing the computational simulation model. A newly-developed computer program needs to be verified through the corresponding benchmark problem before it is used to solve any real geoscience problems. Otherwise, the reliability of the numerical solution obtained from a newly-developed computer program cannot be guaranteed.

1.2.4 Graphical Display of the Numerical Simulation Results

The numerical results obtained from the computer simulation of a geoscience problem are expressed as a large amount of data, which can be viewed using modern technologies of computer graphical display. By comparing the numerical solution with field observations of the geological phenomenon, the correctness of the established conceptual model for the geoscience problem can be tested. Thus the research methodology of computational geoscience is firstly established on the basis of field observations, and then goes through theoretical analysis and computational simulation. Finally the results must be tested through comparison with existing or new field observations. This fundamental research methodology requires that the recognition of a natural phenomenon start from field observations, and be completed through further tests arising from field observations, resulting in a circular iteration.

If the numerical solution is not compatible with the field observations, then the established conceptual model of the problem is questionable and therefore needs to be modified through further refinement of the natural data. On the contrary, if the numerical solution is in accord with the field observations, then the established conceptual model of the geoscience problem is a reasonable interpretation of what may have occurred in nature. In this case, the established conceptual model of the geoscience problem can be further used to investigate the fundamental rules associated with this kind of problem. In this regard, the research methodology of computational geoscience can provide an effective scientific-judging method for solving many controversial problems in the field of geoscience.

1.3 The Contextual Arrangements of this Monograph

In this monograph we use the finite element method, the finite difference method and the particle simulation method as basic numerical methods for dealing with geoscience problems. Based on these three methods, we have developed advanced numerical procedures and algorithms to tackle the large-scale aspects of geoscience problems. The geoscience problems to be considered are closely related to ore body formation and mineralization in hydrothermal systems within the upper crust of the Earth. The arrangements of the forthcoming parts of this monograph are as follows: In Chap. 2, a progressive asymptotic approach procedure is proposed to solve steady-state convective pore-fluid flow problems within the upper crust of the Earth. In combination with the finite element method, this procedure has been applied to simulate convective pore-fluid flow that often plays an important role in ore body formation and mineralization. In Chap. 3, a consistent point-searching interpolation algorithm is proposed to develop a general interface between two commercial computer codes, Fluid Dynamics Analysis Package (FIDAP, Fluid Dynamics International, 1997) and Fast Lagrangian Analysis of Continua (FLAC, Itasca Consulting Group, 1995). With this general interface, the two commercial codes have been used, in an iterative and alternative manner, to solve coupled problems between medium deformation, pore-fluid flow, heat transfer and reactive mass transport processes in hydrothermal systems. In Chap. 4, a term splitting algorithm is developed for dealing with fluid-rock interaction problems that are closely associated with mineral dissolution and precipitation as well as metamorphic processes in fluid-saturated hydrothermal/sedimentary basins of subcritical Zhao numbers. In this case, the chemical dissolution fronts are stable during their propagation within the reactive mass transport system. In contrast, a segregated algorithm is proposed, in Chap. 5, for solving chemical-dissolution front instability problems in fluid-saturated porous rocks of critical and supercritical Zhao numbers. In this situation, the morphological evolution of chemical dissolution fronts in fluid-saturated porous media has been appropriately simulated. In Chap. 6, a decoupling procedure is proposed for simulating fluids mixing, heat transfer and non-equilibrium redox chemical reactions in fluid-saturated porous rocks. The proposed procedure has been applied to investigate the effects of non-equilibrium redox chemical reactions on the mineralization patterns in hydrothermal systems. In Chap. 7, an equivalent source algorithm is presented for simulating thermal and chemical effects of intruded magma solidification problems. This algorithm has been used to simulate effectively and efficiently the thermal and chemical effects of intruded magma in hydrothermal systems. In Chap. 8, the particle simulation method is extended to solve spontaneous crack generation problems in brittle rocks within the upper crust of the Earth. The extended particle method has been applied to simulate spontaneous crack generation associated with faulting and folding in large length-scale geological systems. Finally, some conclusions are given at the end of the monograph.

Chapter 2
A Progressive Asymptotic Approach Procedure for Simulating Steady-State Natural Convective Problems in Fluid-Saturated Porous Media

In a fluid-saturated porous medium, a change in medium temperature may lead to a change in the density of pore-fluid within the medium. This change can be considered as a buoyancy force term in the momentum equation to determine pore-fluid flow in the porous medium using the Oberbeck-Boussinesq approximation model. The momentum equation used to describe pore-fluid flow in a porous medium is usually established using Darcy's law or its extensions. If a fluid-saturated porous medium has the geometry of a horizontal layer, and is heated uniformly from the bottom of the layer, then there exists a temperature difference between the top and bottom boundaries of the layer. Since the positive direction of the temperature gradient due to this temperature difference is opposite to that of the gravity acceleration, there is no natural convection for a small temperature gradient in the porous medium. In this case, heat energy is solely transferred from the high temperature region (the bottom of the horizontal layer) to the low temperature region (the top of the horizontal layer) by thermal conduction. However, if the temperature difference is large enough, it may trigger natural convection in the fluid-saturated porous medium. This problem was first treated analytically by Horton and Rogers (1945) as well as Lapwood (1948), and is often called the Horton-Rogers-Lapwood problem.

This kind of natural convection problem has been found in many geoscience fields. For example, in geoenvironmental engineering, buried nuclear waste and industrial waste in a fluid-saturated porous medium may generate heat and result in a temperature gradient in the vertical direction. If the Rayleigh number, which is directly proportional to the temperature gradient, is equal to or greater than the critical Rayleigh number, natural convection will take place in the porous medium, so that the groundwater may be severely contaminated due to the pore-fluid flow circulation caused by the natural convection. In geophysics, there exists a vertical temperature gradient in the Earth's crust. If this temperature gradient is large enough, it will cause regional natural convection in the Earth's crust. In this situation, the pore-fluid flow circulation due to the natural convection can dissolve soluble minerals in some part of a region and carry them to another part of the region. This is the mineralization problem closely associated with geophysics and geology. Since a natural porous medium is often of a complicated geometry and composed of many different materials, numerical methods are always needed to solve the aforementioned problems.

From the mathematical point of view, the Horton-Rogers-Lapwood problem possesses a bifurcation. The linear stability theory based on the first-order perturbation is commonly used to solve this problem analytically and numerically (Nield 1968, Palm et al. 1972, Caltagirone 1975, 1976, Combarnous and Bories 1975, Buretta and Berman 1976, McKibbin and O'Sullivan 1980, Kaviany 1984, Lebon and Cloot 1986, Pillatsis et al. 1987, Riley and Winters 1989, Islam and Nandakumar 1990, Phillips 1991, Nield and Bejan 1992, Chevalier et al. 1999). However, Joly et al. (1996) pointed out that: "The linear stability theory, in which the nonlinear term of the heat disturbance equation has been neglected, does not describe the amplitude of the resulting convection motion. The computed disturbances are correct only for infinitesimal amplitudes. Indeed, even if the form of convective motion obtained for low supercritical conditions is often quite similar to the critical disturbance, the nonlinear term may produce manifest differences, especially when strong constraints, such as impervious or adiabatic boundaries, are considered." Since it is the amplitude and the form of natural convective motion that significantly affects or dominates the contaminant transport and mineralization in a fluid-saturated porous medium, there is a definite need for including the full nonlinear term of the energy equation in the finite element analysis.

From the finite element analysis point of view, the direct inclusion of the full nonlinear term of the energy equation in the steady-state Horton-Rogers-Lapwood problem would result in a formidable difficulty. The finite element method needs to deal with a highly nonlinear problem and often suffers difficulties in establishing the true non-zero velocity field in a fluid-saturated porous medium because the Horton-Rogers-Lapwood problem always has a zero solution as one possible solution for the velocity field of the pore-fluid. If the velocity field of the pore-fluid used at the beginning of an iteration method is not chosen appropriately, then the resulting finite element solution always tends to zero for the velocity field in a fluid-saturated porous medium. Although this difficulty can be circumvented by turning a steady-state problem into a transient one (Trevisan and Bejan 1987), it is often unnecessary and computationally inefficient to obtain a steady-state solution from solving a transient problem. Therefore, it is highly desirable to develop a numerical procedure to directly solve the steady-state Horton-Rogers-Lapwood problem. For this reason, a progressive asymptotic approach procedure has been developed in recent years (Zhao et al. 1997a, 1998a). The developed progressive asymptotic approach procedure is based on the concept of an asymptotic approach, which was previously and successfully applied to some other fields of the finite element method. For instance, the h-adaptive mesh refinement (Cook et al. 1989) is based on the asymptotic approach concept and can produce a satisfactory solution with the progressive reduction in the size of finite elements used in the analysis. The same asymptotic approach concept was also employed to obtain asymptotic solutions for natural frequencies of vibrating structures in a finite element analysis (Zhao and Steven 1996a, b, c). To solve the steady-state Horton-Rogers-Lapwood problem with the full nonlinear term of the energy equation included in the finite element analysis, the asymptotic approach concept needs to be combined with the finite element method in a different fashion (Zhao et al. 1997a).

2.1 Governing Equations of the Problem

For a two-dimensional fluid-saturated porous medium, if Darcy's law is used to describe pore-fluid flow and the Oberbeck-Boussinesq approximation is employed to describe a change in pore-fluid density due to a change in pore-fluid temperature, the governing equations of a natural convection problem, known as the steady-state Horton-Rogers-Lapwood problem (Nield and Bejan 1992, Zhao et al. 1997a), for incompressible pore-fluid can be expressed as

$$\frac{\partial u}{\partial x} + \frac{\partial v}{\partial y} = 0, \tag{2.1}$$

$$u = \frac{K_x}{\mu}\left(-\frac{\partial P}{\partial x} + \rho_f g_x\right), \tag{2.2}$$

$$v = \frac{K_y}{\mu}\left(-\frac{\partial P}{\partial y} + \rho_f g_y\right), \tag{2.3}$$

$$\rho_{f0} c_p \left(u \frac{\partial T}{\partial x} + v \frac{\partial T}{\partial y}\right) = \lambda_{ex} \frac{\partial^2 T}{\partial x^2} + \lambda_{ey} \frac{\partial^2 T}{\partial y^2}, \tag{2.4}$$

$$\rho_f = \rho_{f0}[1 - \beta_T(T - T_0)], \tag{2.5}$$

$$\lambda_{ex} = \phi \lambda_{fx} + (1-\phi)\lambda_{sx}, \qquad \lambda_{ey} = \phi \lambda_{fy} + (1-\phi)\lambda_{sy}, \tag{2.6}$$

where u and v are the horizontal and vertical velocity components of the pore-fluid in the x and y directions respectively; P is the pore-fluid pressure; T is the temperature of the porous material; K_x and K_y are the permeabilities of the porous material in the x and y directions respectively; μ is the dynamic viscosity of the pore-fluid; ρ_f is the density of the pore-fluid; ρ_{f0} and T_0 are the reference density and temperature; λ_{fx} and λ_{sx} are the thermal conductivities of the pore-fluid and rock mass in the x direction; λ_{fy} and λ_{sy} are the thermal conductivities of the pore-fluid and rock mass in the y direction; c_p is the specific heat of the pore-fluid; g_x and g_y are the gravity acceleration components in the x and y directions; ϕ and β_T are the porosity of the porous material and the thermal volume expansion coefficient of the pore-fluid.

It is noted that Eqs. (2.1), (2.2), (2.3) and (2.4) are derived under the assumption that the porous medium considered is orthotropic, in which the y axis is upward in the vertical direction and coincides with the principal direction of medium permeability as well as that of medium conductivity.

In order to simplify Eqs. (2.1), (2.2), (2.3) and (2.4), the following dimensionless variables are defined:

$$x^* = \frac{x}{H}, \qquad y^* = \frac{y}{H}, \qquad T^* = \frac{T - T_0}{\Delta T}, \qquad (2.7)$$

$$u^* = \frac{H\rho_{f0}c_p}{\lambda_{e0}}u, \qquad v^* = \frac{H\rho_{f0}c_p}{\lambda_{e0}}v, \qquad P^* = \frac{K_h \rho_{f0} c_p}{\mu \lambda_{e0}}(P - P_0), \qquad (2.8)$$

$$K_x^* = \frac{K_x}{K_h}, \qquad K_y^* = \frac{K_y}{K_h}, \qquad \lambda_{ex}^* = \frac{\lambda_{ex}}{\lambda_{e0}}, \qquad \lambda_{ey}^* = \frac{\lambda_{ey}}{\lambda_{e0}}, \qquad (2.9)$$

where x^* and y^* are the dimensionless coordinates; u^* and v^* are the dimensionless velocity components in the x and y directions respectively; P^* and T^* are the dimensionless excess pressure and temperature; K_h is a reference medium permeability coefficient in the horizontal direction; λ_{e0} is a reference conductivity coefficient of the porous medium; $\Delta T = T_{bottom} - T_0$ is the temperature difference between the bottom and top boundaries of the porous medium; H is a reference length and P_0 is the static pore-fluid pressure.

Substituting the above dimensionless variables into Eqs. (2.1), (2.2), (2.3) and (2.4) yields the following dimensionless equations:

$$\frac{\partial u^*}{\partial x^*} + \frac{\partial v^*}{\partial y^*} = 0, \qquad (2.10)$$

$$u^* = K_x^*\left(-\frac{\partial P^*}{\partial x^*} + RaT^*e_1\right), \qquad (2.11)$$

$$v^* = K_y^*\left(-\frac{\partial P^*}{\partial y^*} + RaT^*e_2\right), \qquad (2.12)$$

$$u^*\frac{\partial T^*}{\partial x^*} + v^*\frac{\partial T^*}{\partial y^*} = \lambda_{ex}^*\frac{\partial^2 T^*}{\partial x^{*2}} + \lambda_{ey}^*\frac{\partial^2 T^*}{\partial y^{*2}}, \qquad (2.13)$$

where e is a unit vector and $e = e_1 i + e_2 j$ for a two-dimensional problem; Ra is the Rayleigh number, defined in this particular case as

$$Ra = \frac{(\rho_{f0}c_p)\rho_{f0}g\beta \Delta T K_h H}{\mu \lambda_{e0}}. \qquad (2.14)$$

2.2 Finite Element Formulation of the Problem

By considering the dimensionless velocity, pressure and temperature as basic variables, Eqs. (2.10), (2.11), (2.12) and (2.13) can be discretized using the conventional finite element method (Zienkiewicz 1977, Zhao et al. 1997a). For a typical 4-node quadrilateral element, the velocity, pressure and temperature fields at the elemental level can be expressed as

$$u^*(x^*, y^*) = \boldsymbol{\varphi}^T \mathbf{U}^e, \tag{2.15}$$

$$v^*(x^*, y^*) = \boldsymbol{\varphi}^T \mathbf{V}^e, \tag{2.16}$$

$$P^*(x^*, y^*) = \boldsymbol{\Psi}^T \mathbf{P}^e, \tag{2.17}$$

$$T^*(x^*, y^*) = \boldsymbol{\varphi}^T \mathbf{T}^e, \tag{2.18}$$

where $\mathbf{U}^e$, $\mathbf{V}^e$, $\mathbf{P}^e$ and $\mathbf{T}^e$ are the column vectors of the nodal velocity, excess pressure and temperature of the element; $\boldsymbol{\varphi}$ is the column vector of the interpolation functions for the dimensionless velocity and temperature fields within the element; $\boldsymbol{\Psi}$ is the column vector of the interpolation functions for the excess pressure within the element. For the 4-node quadrilateral element, it is assumed that $\boldsymbol{\varphi}$ is identical to $\boldsymbol{\Psi}$ in the following numerical analysis.

The global coordinate components within the element can be defined as

$$x^* = \mathbf{N}^T \mathbf{X}, \qquad y^* = \mathbf{N}^T \mathbf{Y}, \tag{2.19}$$

where $\mathbf{X}$ and $\mathbf{Y}$ are the column vectors of nodal coordinate components in the x and y directions of the global coordinate system respectively; $\mathbf{N}$ is the column vector of the coordinate mapping function of the element. Based on the isoparametric element concept, the following relationships exist:

$$\mathbf{N}(\xi, \eta) = \boldsymbol{\varphi}(\xi, \eta) = \boldsymbol{\Psi}(\xi, \eta), \tag{2.20}$$

where ξ and η are the local coordinate components of the element.

Using the Galerkin weighted-residual method, Eqs. (2.10), (2.11), (2.12) and (2.13) can be expressed, with consideration of Eqs. (2.15), (2.16), (2.17) and (2.18), as follows:

$$\int_A \boldsymbol{\Psi} \frac{\partial \boldsymbol{\varphi}^T}{\partial x^*} \mathbf{U}^e \, dA + \int_A \boldsymbol{\Psi} \frac{\partial \boldsymbol{\varphi}^T}{\partial y^*} \mathbf{V}^e \, dA = 0, \tag{2.21}$$

$$\int_A \boldsymbol{\varphi}\boldsymbol{\varphi}^T \mathbf{U}^e \, dA + \int_A \boldsymbol{\varphi} K_x^* \frac{\partial \boldsymbol{\Psi}^T}{\partial x^*} \mathbf{P}^e \, dA + \int_A \boldsymbol{\varphi} K_x^* Ra \boldsymbol{\varphi}^T \mathbf{T}^e \, e_1 dA = 0, \tag{2.22}$$

$$\int_A \varphi \varphi^T \mathbf{V}^e \, dA + \int_A \varphi K_y^* \frac{\partial \mathbf{\Psi}^T}{\partial y^*} \mathbf{P}^e \, dA + \int_A \varphi K_y^* Ra \varphi^T \mathbf{T}^e e_2 \, dA = 0, \quad (2.23)$$

$$\int_A \varphi u^* \frac{\partial \varphi^T}{\partial x^*} \mathbf{T}^e \, dA + \int_A \varphi v^* \frac{\partial \varphi^T}{\partial y^*} \mathbf{T}^e \, dA - \int_A \varphi \lambda_{ex}^* \frac{\partial^2 \varphi^T}{\partial x^{*2}} \mathbf{T}^e \, dA - \int_A \varphi \lambda_{ey}^* \frac{\partial^2 \varphi^T}{\partial y^{*2}} \mathbf{T}^e \, dA = 0. \quad (2.24)$$

Using the Green-Gauss theorem and the technique of integration by parts, the terms involving the second derivatives in Eq. (2.24) can be rewritten as

$$\int_A \varphi \lambda_{ex}^* \frac{\partial^2 \varphi^T}{\partial x^{*2}} \mathbf{T}^e \, dA = -\int_A \frac{\partial \varphi}{\partial x^*} \lambda_{ex}^* \frac{\partial \varphi^T}{\partial x^*} \mathbf{T}^e \, dA + \int_S \varphi q_x^* n_x \, dS = 0, \quad (2.25)$$

$$\int_A \varphi \lambda_{ey}^* \frac{\partial^2 \varphi^T}{\partial y^{*2}} \mathbf{T}^e \, dA = -\int_A \frac{\partial \varphi}{\partial y^*} \lambda_{ey}^* \frac{\partial \varphi^T}{\partial y^*} \mathbf{T}^e \, dA + \int_S \varphi q_y^* n_y \, dS = 0, \quad (2.26)$$

where q_x^* and q_y^* are the dimensionless heat fluxes on the element boundary of a unit normal vector, $\mathbf{n}$; A and S are the area and boundary length of the element.

Note that Eqs. (2.21), (2.22), (2.23) and (2.24) can be expressed in a matrix form as follows:

$$\begin{bmatrix} \mathbf{M}^e & 0 & -\mathbf{B}_x^e & -\mathbf{A}_x^e \\ 0 & \mathbf{M}^e & -\mathbf{B}_y^e & -\mathbf{A}_y^e \\ 0 & 0 & \mathbf{E}^e & 0 \\ \mathbf{C}_x^e & \mathbf{C}_y^e & 0 & 0 \end{bmatrix} \begin{Bmatrix} \mathbf{U}^e \\ \mathbf{V}^e \\ \mathbf{T}^e \\ \mathbf{P}^e \end{Bmatrix} = \begin{Bmatrix} \mathbf{F}_x^e \\ \mathbf{F}_y^e \\ \mathbf{G}^e \\ 0 \end{Bmatrix}, \quad (2.27)$$

where $\mathbf{U}^e$ and $\mathbf{V}^e$ are the nodal dimensionless velocity vectors of the element in the x and y directions respectively; $\mathbf{T}^e$ and $\mathbf{P}^e$ are the nodal dimensionless temperature and pressure vectors of the element; $\mathbf{A}_x^e$, $\mathbf{A}_y^e$, $\mathbf{B}_x^e$, $\mathbf{B}_y^e$, $\mathbf{C}_x^e$, $\mathbf{C}_y^e$, $\mathbf{E}^e$ and $\mathbf{M}^e$ are the property matrices of the element; $\mathbf{F}_x^e$, $\mathbf{F}_y^e$ and $\mathbf{G}^e$ are the dimensionless nodal load vectors due to the dimensionless stress and heat flux on the boundary of the element. These matrices and vectors can be derived and expressed as follows:

$$\mathbf{A}_x^e = \int_A \frac{\partial \varphi}{\partial x^*} K_x^* \mathbf{\Psi}^T \, dA, \qquad \mathbf{B}_x^e = \int_A \varphi K_x^* Ra \varphi^T e_1 \, dA, \qquad \mathbf{C}_x^e = \int_A \mathbf{\Psi} \frac{\partial \varphi^T}{\partial x^*} dA, \quad (2.28)$$

$$\mathbf{A}_y^e = \int_A \frac{\partial \varphi}{\partial y^*} K_y^* \mathbf{\Psi}^T \, dA, \qquad \mathbf{B}_y^e = \int_A \varphi K_y^* Ra \varphi^T e_2 \, dA, \qquad \mathbf{C}_y^e = \int_A \mathbf{\Psi} \frac{\partial \varphi^T}{\partial y^*} dA, \quad (2.29)$$

$$\mathbf{D}_x^e(u^*) = \int_A \varphi u^* \frac{\partial \varphi^T}{\partial x^*} dA, \qquad \mathbf{L}_x^e = \int_A \frac{\partial \varphi}{\partial x^*} \lambda_{ex}^* \frac{\partial \varphi}{\partial x^*} dA, \qquad \mathbf{F}_x^e = \int_S \sigma_x^* \varphi \, dS, \quad (2.30)$$

2.2 Finite Element Formulation of the Problem

$$\mathbf{D}_y^e(v^*) = \int_A \varphi v^* \frac{\partial \varphi^T}{\partial x^*} dA, \quad \mathbf{L}_y^e = \int_A \frac{\partial \varphi}{\partial y^*} \lambda_{ey}^* \frac{\partial \varphi}{\partial y^*} dA, \quad \mathbf{F}_y^e = \int_S \sigma_y^* \varphi dS, \quad (2.31)$$

$$\mathbf{E}^e = \mathbf{D}_x^e(u^*) + \mathbf{D}_y^e(v^*) + \mathbf{L}_x^e + \mathbf{L}_y^e, \quad \mathbf{G}^e = -\int_S q^* \varphi dS, \quad (2.32)$$

$$\mathbf{M}^e = \int_A \varphi \varphi^T dA, \quad q^* = \frac{H}{\Delta T \lambda_{e0}} q, \quad \sigma^* = \frac{K_h \rho_{f0} c_p}{\mu \lambda_{e0}} \sigma, \quad (2.33)$$

where φ is the shape function vector for the temperature and velocity components of the element; Ψ is the shape function vector for the pressure of the element; σ and q are the stress and heat flux on the boundary of the element; A and S are the area and boundary length of the element.

It is noted that since the full nonlinear term of the energy equation in the Horton-Rogers-Lapwood problem is considered in the finite element analysis, matrix $\mathbf{E}^e$ is dependent on the velocity components of the element. Thus, a prediction for the initial velocities of an element is needed to have this matrix evaluated. This is the main motivation for proposing a progressive asymptotic approach procedure in the next section.

From the penalty finite element approach (Zienkiewicz 1977), the following equation exists:

$$\mathbf{C}_x^e \mathbf{U}^e + \mathbf{C}_y^e \mathbf{V}^e = -\varepsilon \mathbf{M}_p \mathbf{P}^e. \quad (2.34)$$

Equation (2.34) can be rewritten as

$$\mathbf{P}^e = -\frac{1}{\varepsilon} \mathbf{M}_p^{-1} (\mathbf{C}_x^e \mathbf{U}^e + \mathbf{C}_y^e \mathbf{V}^e). \quad (2.35)$$

Substituting Eq. (2.35) into Eq. (2.27) yields the following equation in the elemental level:

$$\begin{bmatrix} \mathbf{Q}^e & -\mathbf{B}^e \\ 0 & \mathbf{E}^e \end{bmatrix} \begin{Bmatrix} \mathbf{U}_F^e \\ \mathbf{T}^e \end{Bmatrix} = \begin{Bmatrix} \mathbf{F}^e \\ \mathbf{G}^e \end{Bmatrix}, \quad (2.36)$$

where

$$\mathbf{Q}^e = \overline{\mathbf{M}}^e + \frac{1}{\varepsilon} \mathbf{A}^e (\mathbf{M}_p^e)^{-1} (\mathbf{C}^e)^T, \quad (2.37)$$

$$\overline{\mathbf{M}}^e = \begin{bmatrix} \mathbf{M}^e & 0 \\ 0 & \mathbf{M}^e \end{bmatrix}, \quad \mathbf{U}_F^e = \begin{Bmatrix} \mathbf{U}^e \\ \mathbf{V}^e \end{Bmatrix}, \quad \mathbf{F}^e = \begin{Bmatrix} \mathbf{F}_x^e \\ \mathbf{F}_y^e \end{Bmatrix}, \quad (2.38)$$

$$\mathbf{B}^e = \begin{Bmatrix} \mathbf{B}_x^e \\ \mathbf{B}_y^e \end{Bmatrix}, \quad \mathbf{A}^e = \begin{Bmatrix} \mathbf{A}_x^e \\ \mathbf{A}_y^e \end{Bmatrix}, \quad \mathbf{C}^e = \begin{Bmatrix} \mathbf{C}_x^e \\ \mathbf{C}_y^e \end{Bmatrix}, \quad (2.39)$$

$$\mathbf{M}_p^e = \int_A \boldsymbol{\Psi}\boldsymbol{\Psi}^T dA. \tag{2.40}$$

It needs to be pointed out that ε is a penalty parameter in Eq. (2.37). For the purpose of obtaining an accurate solution, this parameter must be chosen small enough to approximate fluid incompressibility well, but large enough to prevent the resulting matrix problem from becoming too ill-conditioned to solve.

By assembling all elements in a system, the finite element equation of the system can be expressed in a matrix form as

$$\begin{bmatrix} \mathbf{Q} & -\mathbf{B} \\ \mathbf{0} & \mathbf{E}(\mathbf{U}) \end{bmatrix} \begin{Bmatrix} \mathbf{U_F} \\ \mathbf{T} \end{Bmatrix} = \begin{Bmatrix} \mathbf{F} \\ \mathbf{G} \end{Bmatrix}, \tag{2.41}$$

where $\mathbf{Q}$, $\mathbf{B}$ and $\mathbf{E}$ are global property matrices of the system; $\mathbf{U_F}$ and $\mathbf{T}$ are global nodal velocity and temperature vectors of the system; $\mathbf{F}$ and $\mathbf{G}$ are global nodal load vectors of the system. Since Equation (2.41) is nonlinear, either the successive substitution method or the Newton-Raphson method can be used to solve this equation.

2.3 The Progressive Asymptotic Approach Procedure for Solving Steady-State Natural Convection Problems in Fluid-Saturated Porous Media

To solve the steady-state Horton-Rogers-Lapwood problem with the full nonlinear term of the energy equation included in the finite element analysis, the asymptotic approach concept (Cook et al. 1989, Zhao and Steven 1996a, b, c) needs to be used in a progressive fashion (Zhao et al. 1997a). If the gravity acceleration is assumed to tilt at a small angle, α, in the Horton -Rogers-Lapwood problem, then a non-zero velocity field in a fluid-saturated porous medium may be found using the finite element method. The resulting non-zero velocity field can be used as the initial velocity field of the pore-fluid to solve the original Horton-Rogers-Lapwood problem with the tilted small angle being zero. Thus, two kinds of problems need to be progressively solved in the finite element analysis. One is the modified Horton-Rogers-Lapwood problem, in which the gravity acceleration is tilted a small angle, and another is the original Horton-Rogers-Lapwood problem. This forms two basic steps of the progressive asymptotic approach procedure. Clearly, the basic idea behind the progressive asymptotic approach procedure is that when the small angle tilted by the gravity acceleration approaches zero, the modified Horton-Rogers-Lapwood problem asymptotically approaches the original one and as a result, a solution to the original Horton-Rogers-Lapwood problem can be obtained.

Based on the basic idea behind the progressive asymptotic approach procedure, the key issue of obtaining a non-zero pore-fluid flow solution for the Horton-Rogers-Lapwood problem is to choose the initial velocity field of pore-fluid correctly. If the initial velocity field is not correctly chosen, the finite element method will lead to

2.3 Solving Steady-State Natural Convection Problems in Fluid-Saturated Porous Media

a zero pore-fluid flow solution for natural convection of pore-fluid, even though the Rayleigh number is high enough to drive the occurrence of natural convection in a fluid-saturated porous medium. In order to overcome this difficulty, a modified Horton-Rogers-Lapwood problem, in which the gravity acceleration is assumed to tilt a small angle α, needs to be solved. Supposing the original Horton-Rogers-Lapwood problem has a Rayleigh number (Ra) and that the non-zero solution for the modified Horton-Rogers-Lapwood problem is $S(Ra, \alpha)$, it is possible to find a non-zero solution for the original Horton-Rogers-Lapwood problem by taking a limit of $S(Ra, \alpha)$ when α approaches zero. This process can be mathematically expressed as follows:

$$\lim_{\alpha \to 0} S(Ra, \alpha) = S(Ra, 0), \quad (2.42)$$

where $S(Ra, 0)$ is a solution for the original Horton-Rogers-Lapwood problem; $S(Ra, \alpha)$ is the solution for the modified Horton-Rogers-Lapwood problem; S is any variable to be solved in the original Horton-Rogers-Lapwood problem.

It is noted that in theory, if $S(Ra, \alpha)$ could be expressed as a function of α explicitly, $S(Ra, 0)$ would follow immediately. However, in practice, it is necessary to find out $S(Ra, 0)$ numerically since it is very difficult and often impossible to express $S(Ra, \alpha)$ in an explicit manner. Thus, the question which must be answered is how to choose α so as to obtain an accurate non-zero solution, $S(Ra, 0)$. From the theoretical point of view, it is desirable to choose α as small as possible. The reason for this is that the smaller the value of α, the closer the characteristic of $S(Ra, \alpha)$ to that of $S(Ra, 0)$. This enables a more accurate solution $S(Ra, 0)$ to be obtained in the computation. From the finite element analysis point of view, α cannot be chosen too small because the smaller the value of α, the more sensitive the solution $S(Ra, \alpha)$ to the initial velocity field of pore-fluid. As a result, a very small α usually leads to a zero velocity field due to any inappropriate choice for the initial velocity field of pore-fluid. To avoid this phenomenon, α should be chosen big enough to eliminate the strong dependence of $S(Ra, \alpha)$ on the initial velocity field of pore-fluid. For the purpose of using a big value of α and keeping the final solution $S(Ra, 0)$ of good accuracy in the finite element analysis, $S(Ra, \alpha)$ needs to approach $S(Ra, 0)$ in a progressive asymptotic manner, as clearly shown in Fig. 2.1. This leads to the following processes mathematically:

$$\lim_{\alpha_i \to \alpha_{i+1}} S(Ra, \alpha_i) = S(Ra, \alpha_{i+1}) \quad (i = 1, 2, \ldots, n-1), \quad (2.43)$$

$$\lim_{\alpha_n \to 0} S(Ra, \alpha_n) = S(Ra, 0), \quad (2.44)$$

$$\alpha_1 = \alpha, \qquad \alpha_{i+1} = \frac{1}{R}\alpha_i, \quad (2.45)$$

where n is the total step number for α approaching zero; R is the rate of α_i approaching α_{i+1}. Generally, the values of α, n and R are dependent on the nature of a problem to be analysed.

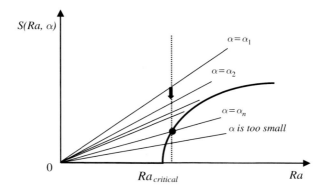

Fig. 2.1 The basic concept of the progressive asymptotic approach procedure

For solving the steady-state Horton-Rogers-Lapwood problem using the progressive asymptotic approach procedure associated with the finite element method, numerical experience has shown that $1° \leq \alpha \leq 5°$, $5 \leq R \leq 10$ and $1 \leq n \leq 2$ leads to acceptable solutions. Therefore, for α in the range of 1–5° and R in the range of 5–10, $S(Ra, \alpha)$ can asymptotically approach $S(Ra, 0)$ in one step or two steps. This indicates the efficiency of the present procedure.

2.4 Derivation of Analytical Solution to a Benchmark Problem

In order to verify the applicability of the progressive asymptotic approach procedure for solving the Horton-Rogers-Lapwood convection problem, an analytical solution is needed for a benchmark problem, the geometry and boundary conditions of which can be exactly modelled by the finite element method. Although the existing solutions (Phillips 1991, Nield and Bejan 1992) for a horizontal layer in porous media can be used to check the accuracy of a finite element solution within a square box with appropriate boundary conditions, it is highly desirable to examine the progressive asymptotic approach procedure as extensively as possible. For this purpose, a benchmark problem of any rectangular geometry is constructed and shown in Fig. 2.2. Without losing generality, the dimensionless governing equations given in Eqs. (2.10), (2.11), (2.12) and (2.13) are considered in this section. The boundary conditions of the benchmark problem are expressed using the dimensionless variables as follows:

$$u^* = 0, \quad \frac{\partial T^*}{\partial x^*} = 0 \quad (\text{at } x^* = 0 \text{ and } x^* = L^*), \quad (2.46)$$

$$v^* = 0, \quad T^* = 1 \quad (\text{at } y^* = 0), \quad (2.47)$$

$$v^* = 0, \quad T^* = 0 \quad (\text{at } y^* = 1), \quad (2.48)$$

2.4 Derivation of Analytical Solution to a Benchmark Problem

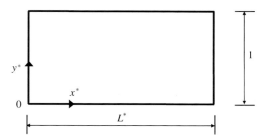

Fig. 2.2 Geometry of a benchmark problem

where L^* is a dimensionless length in the horizontal direction and $L^* = L/H$, in which L is the real length of the problem domain in the horizontal direction.

For ease of deriving an analytical solution to the benchmark problem, it is assumed that the porous medium under consideration is fluid-saturated and isotropic. This means that $K_x = K_y = K_h$ and $\lambda_{ex} = \lambda_{ey} = \lambda_{e0}$. As a result, Eqs. (2.10), (2.11), (2.12) and (2.13) can be further simplified as follows:

$$\frac{\partial u^*}{\partial x^*} + \frac{\partial v^*}{\partial y^*} = 0, \tag{2.49}$$

$$u^* = -\frac{\partial P^*}{\partial x^*} + RaT^*e_1, \tag{2.50}$$

$$v^* = -\frac{\partial P^*}{\partial y^*} + RaT^*e_2, \tag{2.51}$$

$$u^*\frac{\partial T^*}{\partial x^*} + v^*\frac{\partial T^*}{\partial y^*} = \frac{\partial^2 T^*}{\partial x^{*2}} + \frac{\partial^2 T^*}{\partial y^{*2}}. \tag{2.52}$$

Using the linearization procedure for temperature gradient and a dimensionless stream function Ψ simultaneously, Eqs. (2.49), (2.50), (2.51) and (2.52) are reduced to the following two equations:

$$\frac{\partial^2 \Psi}{\partial x^{*2}} + \frac{\partial^2 \Psi}{\partial y^{*2}} = -Ra\frac{\partial T^*}{\partial x^*}, \tag{2.53}$$

$$\frac{\partial \Psi}{\partial x^*} = \frac{\partial^2 T^*}{\partial x^{*2}} + \frac{\partial^2 T^*}{\partial y^{*2}}. \tag{2.54}$$

Since Eqs. (2.53) and (2.54) are linear, solutions to Ψ and T^* are of the following forms:

$$\Psi = f(y^*)\sin\left(q\frac{x^*}{L^*}\right) \quad (q = m\pi, m = 1, 2, 3, \ldots\ldots), \tag{2.55}$$

$$T^* = \theta(y^*)\cos\left(q\frac{x^*}{L^*}\right) + (1 - y^*) \quad (q = m\pi, m = 1, 2, 3, \ldots). \quad (2.56)$$

Substituting Eqs. (2.55) and (2.56) into Eqs. (2.53) and (2.54) yields the following equations:

$$f''(y^*) - \left(\frac{q}{L^*}\right)^2 f(y^*) = \frac{q}{L^*} Ra\theta(y^*), \quad (2.57)$$

$$\frac{q}{L^*} f(y^*) = -\left(\frac{q}{L^*}\right)^2 \theta(y^*) + \theta''(y^*). \quad (2.58)$$

Combining Eqs. (2.57) and (2.58) leads to an equation containing $f(y^*)$ only:

$$f^{IV}(y^*) - 2\left(\frac{q}{L^*}\right)^2 f''(y^*) - \left(\frac{q}{L^*}\right)^2 \left[Ra - \left(\frac{q}{L^*}\right)^2\right] f(y^*) = 0. \quad (2.59)$$

It is immediately noted that Equation (2.59) is a linear, homogeneous ordinary differential equation so that it has a zero trivial solution. For the purpose of finding out a non-zero solution, it is noted that the non-zero solution satisfying both Equation (2.59) and the boundary conditions in Eqs. (2.46), (2.47) and (2.48) can be expressed as

$$f(y^*) = \sin(ry^*) \quad (r = n\pi, n = 1, 2, 3, \ldots). \quad (2.60)$$

Using this equation, the condition under which the non-zero solution exists for Eq. (2.59) is derived and expressed as

$$Ra = \left(\frac{L^*}{q}r^2 + \frac{q}{L^*}\right)^2 = \left(\frac{n^2}{m}L^* + \frac{m}{L^*}\right)^2 \pi^2 \quad (2.61)$$
$$(m = 1, 2, 3, \ldots, n = 1, 2, 3, \ldots).$$

It can be observed from Eq. (2.61) that in the case of L^* being an integer, the minimum Rayleigh number is $4\pi^2$, which occurs when $n = 1$ and $m = L^*$. However, if L^* is not an integer, the minimum Rayleigh number is $(L^* + 1/L^*)^2\pi^2$, which occurs when $m = 1$ and $n = 1$. Since the minimum Rayleigh number determines the onset of natural convection in a fluid-saturated porous medium for the Horton-Rogers-Lapwood problem, it is often labelled as the critical Rayleigh number, $Ra_{critical}$.

For this benchmark problem, the mode shapes for the stream function and related dimensionless variables corresponding to the critical Rayleigh number can be derived and expressed as follows:

$$\Psi = C_1 \sin\left(\frac{m\pi}{L^*}x^*\right)\sin(n\pi y^*), \quad (2.62)$$

2.5 Verification of the Proposed Progressive Asymptotic Approach

$$u^* = n\pi C_1 \sin\left(\frac{m\pi}{L^*}x^*\right)\cos(n\pi y^*), \qquad (2.63)$$

$$v^* = -\frac{m\pi}{L^*}C_1 \cos\left(\frac{m\pi}{L^*}x^*\right)\sin(n\pi y^*), \qquad (2.64)$$

$$T^* = -\frac{C_1}{\sqrt{Ra_{critical}}} \cos\left(\frac{m\pi}{L^*}x^*\right)\sin(n\pi y^*) + (1 - y^*), \qquad (2.65)$$

$$P^* = \frac{nL^*}{m}C_1 \cos\left(\frac{m\pi}{L^*}x^*\right)\cos(n\pi y^*) - \frac{Ra_{critical}}{2}(1 - y^*)^2 + C_2, \qquad (2.66)$$

where the values of m, n and $Ra_{critical}$ are dependent on whether L^* is an integer or not; C_1 is a non-zero constant and C_2 is an arbitrary constant. It is interesting to note that since $Ra_{critical}$ is a function of L^*, it can vary with a non-integer L^*. This implies that if rectangular valleys are filled with porous media, they may have different critical Rayleigh numbers when their ratios of length to height are different.

2.5 Verification of the Proposed Progressive Asymptotic Approach Procedure Associated with Finite Element Analysis

Using the analytical solution derived for a benchmark problem in the last section, the proposed progressive asymptotic approach procedure associated with the finite element analysis for solving the Horton-Rogers-Lapwood problem in a fluid-saturated porous medium is verified in this section. A rectangular domain of $L^* = 1.5$ is considered in the calculation. The critical Rayleigh number for the test problem considered is $169\pi^2/36$. As shown in Fig. 2.3, the problem domain is discretized into

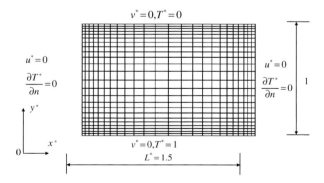

Fig. 2.3 Finite element mesh for the benchmark problem

864 nine-node quadrilateral elements of 3577 nodes in total. The mesh gradation technique, which enables the region in the vicinity of problem boundaries to be modelled using finite elements of small sizes, has been employed to increase the solution accuracy in this region. The following parameters associated with the progressive asymptotic approach procedure are used in the calculation: $\alpha = 5^o$, $n = 2$ and $R = 5$.

Figures 2.4, 2.5, 2.6 and 2.7 show the comparison of numerical solutions with analytical ones for dimensionless velocity, stream function, temperature and pressure modes respectively. In these figures, the plots above are analytical solutions, whereas the plots below are numerical solutions for the problem. It is observed from these results that the numerical solutions from the progressive asymptotic approach procedure associated with the finite element method are in good agreement with the analytical solutions. Compared with the analytical solutions, the maximum error in the numerical solutions is less than 2%. This demonstrates the usefulness of the present progressive asymptotic approach procedure when it is used to solve the steady-state Horton-Rogers-Lapwood problems.

At this point, there is a need to explain why both the analytical and the numerical solutions for the pore-fluid flow are non-symmetric, although the geometry and boundary conditions for the problem are symmetric. As stated previously, the Horton-Rogers-Lapwood problem belongs mathematically to a bifurcation problem. The trivial solution for the pore-fluid flow of the problem is zero. That is to say, if the Rayleigh number of the problem is less than the critical Rayleigh number, the solution resulting from any small disturbance or perturbation converges to the trivial

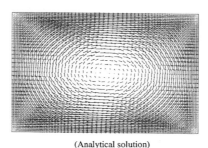

(Analytical solution)

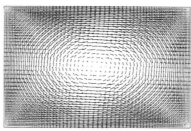

(Numerical solution)

Fig. 2.4 Comparison of numerical solution with analytical solution (Dimensionless velocity)

2.5 Verification of the Proposed Progressive Asymptotic Approach

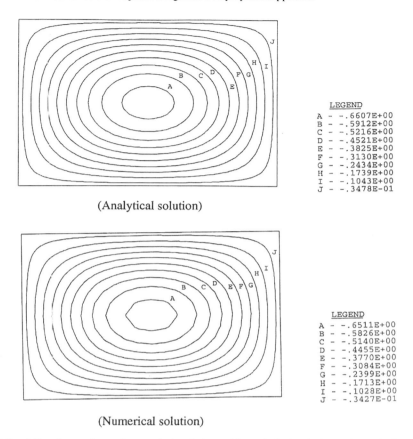

Fig. 2.5 Comparison of numerical solution with analytical solution (Dimensionless stream function)

solution. In this case, the solution for the pore-fluid flow is zero (and, of course, symmetric) and the system is in a stable state. However, if the Rayleigh number of the problem is equal to or greater than the critical Rayleigh number, the solution resulting from any small disturbance or perturbation may lead to a non-trivial solution. In this situation, the solution for the pore-fluid flow is non-zero and the system is in an unstable state. Since the main purpose of this study is to find out the non-trivial solution for problems having a high Rayleigh number, $Ra \geq Ra_{critical}$, a small disturbance or perturbation needs to be applied to the system at the beginning of a computation. This is why gravity is firstly tilted a small angle away from vertical and then gradually approaches and is finally restored to vertical in the proposed progressive asymptotic approach procedure. It is the small perturbation that makes the non-trivial solution non-symmetric, even though the system considered is symmetric. In addition, as addressed in Sect. 2.3, the solution dependence on the amplitude of the initially-tilted small angle can be avoided by making this angle approach zero

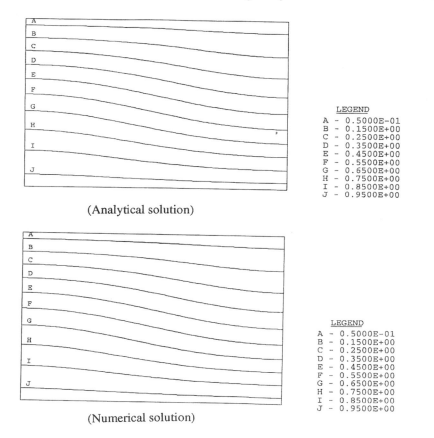

Fig. 2.6 Comparison of numerical solution with analytical solution (Dimensionless temperature)

in a progressive asymptotic manner. However, since there are two possible non-symmetric solutions for convective pore-fluid flow, namely a clockwise convective flow and an anti-clockwise convective flow, the solution dependence on the direction of the initially-tilted small angle cannot be avoided.

2.6 Application of the Progressive Asymptotic Approach Procedure Associated with Finite Element Analysis

2.6.1 Two-Dimensional Convective Pore-Fluid Flow Problems

The present progressive asymptotic approach procedure is employed to investigate the effect of basin shapes on natural convection in a fluid-saturated porous medium when it is heated from below. Three different basin shapes having square,

2.6 Application of the Progressive Asymptotic Approach Procedure

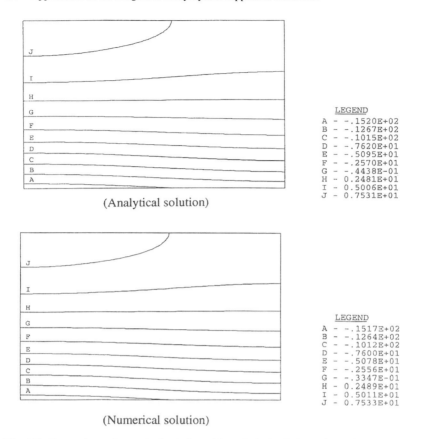

Fig. 2.7 Comparison of numerical solution with analytical solution (Dimensionless pressure)

rectangular and trapezoidal geometries, which are filled with fluid-saturated porous media, are considered in the analysis. For the rectangular basin, the ratio of width to height is 1.5. For the trapezoidal basin, the ratios of top width to height and bottom width to height are 2 and 1 respectively. In order to reflect the anisotropic behaviour of the porous media, the medium permeability in the horizontal direction is assumed to be three times that in the vertical direction. As shown in Fig. 2.8, all three basins are discretized into 484 nine-node quadrilateral elements of 2041 nodes in total. The boundary conditions of the problems are also shown in Fig. 2.8, in which n is the normal direction of a boundary. Two Rayleigh numbers, namely $Ra = 80$ and $Ra = 400$, are used to examine the effect of the Rayleigh number on natural convection in a fluid-saturated porous medium. The same parameters as used in the above model verification examples have been used here for the progressive asymptotic approach procedure.

Figure 2.9 shows the dimensionless velocity distribution for the three different basins, whereas Figs. 2.10 and 2.11 show the dimensionless streamline contours

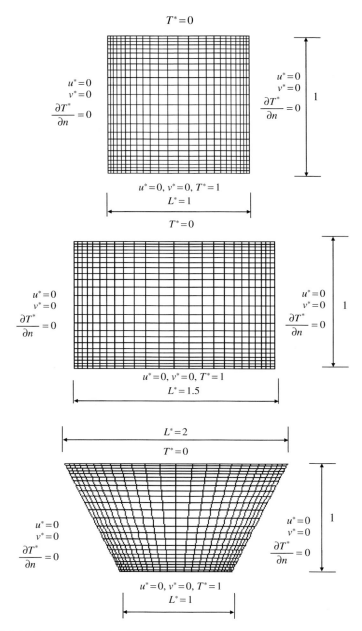

Fig. 2.8 Finite element meshes for three different basins shapes

2.6 Application of the Progressive Asymptotic Approach Procedure

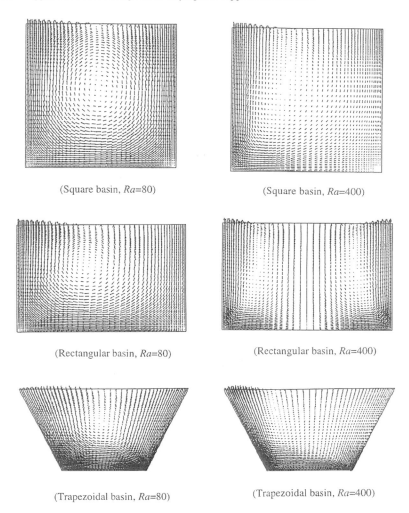

Fig. 2.9 Dimensionless velocity distribution for different basins

due to different basin shapes for $Ra = 80$ and $Ra = 400$ respectively. It is obvious that different basin shapes have a considerable effect on the patterns of convective flow in the fluid-saturated porous medium, especially in the case of higher Rayleigh numbers. Apart from notable differences in velocity distribution patterns, maximum velocity amplitudes for three different basins are also significantly different. For instance, in the case of $Ra = 80$, the maximum amplitudes of dimensionless velocities are 5.29, 8.93 and 11.66 for square, rectangular and trapezoidal basins respectively. This fact indicates that different basin shapes may affect the contaminant transport or mineralization processes in a fluid-saturated porous medium once natural convection is initiated in the medium.

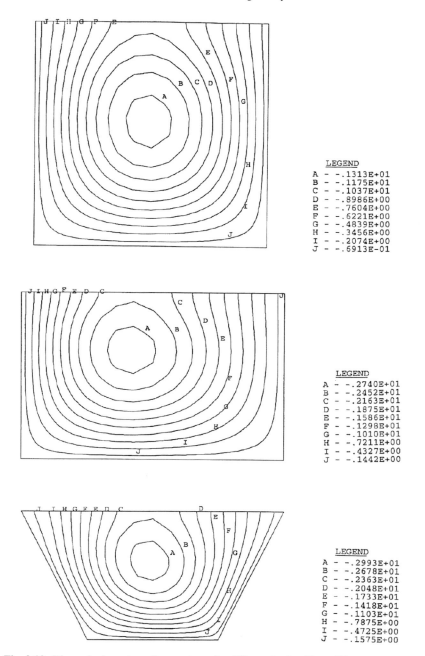

Fig. 2.10 Dimensionless streamline contours for different basins ($Ra = 80$)

2.6 Application of the Progressive Asymptotic Approach Procedure

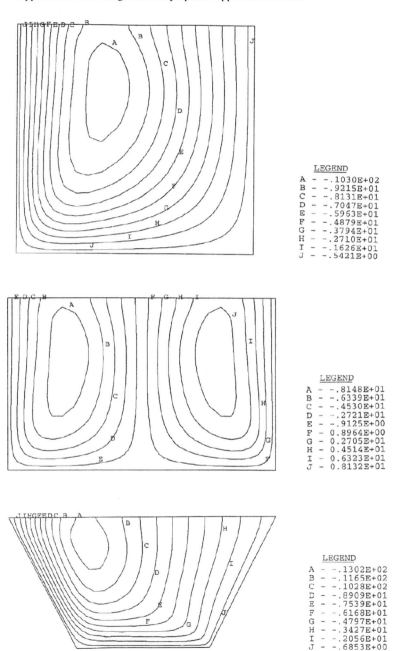

Fig. 2.11 Dimensionless streamline contours for different basins ($Ra = 400$)

2.6.2 Three-Dimensional Convective Pore-Fluid Flow Problems

The proposed progressive asymptotic approach procedure can be straightforwardly extended to the simulation of three-dimensional convective pore-fluid flow problems in fluid-saturated porous media (Zhao et al. 2001a, 2003a). Since the streamline function is not available for three dimensional fluid flow problems, it is necessary to use the particle tracking technique to show particle movements in three dimensional fluid flow systems. In the particle tracking technique, a fundamental problem, which needs to be solved effectively, is as follows. If the initial (known) location of a particle is point A (x_A, y_A, z_A), where x_A, y_A and z_A are the coordinate components of point A in the x, y and z directions of a global coordinate system, then we need to determine where the new location (i.e. point A') of this particle is after a given time interval, Δt. Clearly, if the velocity of the particle at point A is known, then the coordinate components of point A' in the global coordinate system can be approximately determined for a small Δt as follows:

$$x_{A'} = x_A + u_A \Delta t, \tag{2.67}$$

$$y_{A'} = y_A + v_A \Delta t, \tag{2.68}$$

$$z_{A'} = z_A + w_A \Delta t, \tag{2.69}$$

where $x_{A'}$, $y_{A'}$ and $z_{A'}$ are the coordinate components of point A' in the x, y and z directions of the global coordinate system; u_A, v_A and w_A are the velocity components of point A in the x, y and z directions of the global coordinate system, respectively.

In general cases, the location of point A is not coincident with the nodal points in a finite element analysis so that the consistent interpolation of the finite element solution is needed to determine the velocity components at this point. For this purpose, it is essential to find the coordinate components of point A in the local coordinate system from the following equations for an isoparametric finite element.

$$x_A = \sum_{i=1}^{n} \phi_i(\xi_A, \eta_A, \zeta_A) x_i, \tag{2.70}$$

$$y_A = \sum_{i=1}^{n} \phi_i(\xi_A, \eta_A, \zeta_A) y_i, \tag{2.71}$$

$$z_A = \sum_{i=1}^{n} \phi_i(\xi_A, \eta_A, \zeta_A) z_i, \tag{2.72}$$

where x_A, y_A and z_A are the coordinate components of point A in the x, y and z directions of the global coordinate system; ξ_A, η_A and ζ_A are three coordinate components

2.6 Application of the Progressive Asymptotic Approach Procedure

of point A in the ξ, η and ζ directions of a local coordinate system; x_i, y_i and z_i are the coordinate components of nodal point i in the x, y and z directions of the global coordinate system; n is the total nodal number of the element containing point A; ϕ_i is the interpolation function of node i in the element containing point A.

In Eqs. (2.70), (2.71) and (2.72), the coordinate components of point A in the global coordinate system are known, so that the coordinate components of this point in the local system can be determined using any inverse mapping technique (Zhao et al. 1999f). Once the coordinate components of point A in the global coordinate system are determined, the velocity components of point A in the global system can be straightforwardly calculated as follows:

$$u_A = \sum_{i=1}^{n} \phi_i(\xi_A, \eta_A, \zeta_A) u_i \qquad (2.73)$$

$$v_A = \sum_{i=1}^{n} \phi_i(\xi_A, \eta_A, \zeta_A) v_i \qquad (2.74)$$

$$w_A = \sum_{i=1}^{n} \phi_i(\xi_A, \eta_A, \zeta_A) w_i, \qquad (2.75)$$

where u_A, v_A and w_A are the velocity components of point A in the x, y and z directions of the global coordinate system; u_i, v_i and w_i are the velocity components of nodal point i in the x, y and z directions of the global coordinate system, respectively.

The above-mentioned process indicates that in the finite element analysis of fluid flow problems, the trajectory of any given particle can be calculated using the nodal coordinate and velocity components, which are fundamental quantities and therefore available in the finite element analysis.

To demonstrate the applicability of the progressive asymptotic approach procedure for simulating convective pore-fluid flow in three dimensional situations, the example considered in this section is a cubic box of $10 \times 10 \times 10\,km^3$ in size. This box is filled with pore-fluid saturated porous rock, which is a part of the upper crust of the Earth. In order to simulate geothermal conditions in geology, the bottom of the box is assumed to be hotter than the top of the box. This means that the pore-fluid saturated porous rock is uniformly heated from below. For the system considered here, the classical analysis (Phillips 1991, Nield and Bejan 1992, Zhao et al. 1997a) indicates that the convective flow is possible when the Rayleigh number of the system is either critical or supercritical. For this reason, the parameters and properties of the system are deliberately selected in such a way that the Rayleigh number of the system is supercritical.

Figure 2.12 shows the finite element mesh of 8000 cubic elements for the three dimensional convective flow problem. For the purpose of investigating the perturbation direction on the pattern of convective flow, two cases are considered in the following computations. In the first case, the perturbation of gravity is applied in the x-z plane only. This means that the problem is axisymmetrical about the y axis so that

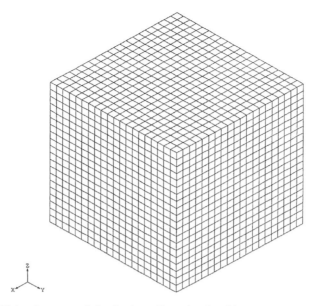

Fig. 2.12 Finite element mesh for the three-dimensional problem

the problem can be degenerated into a two dimensional problem, from the mathematical and analytical points of view. Since the solutions for the axisymmetrical convective flow problem are available (Nield and Bejan 1992, Zhao et al. 1997a), the numerical methods used in this study can be verified by comparing the related analytical solutions with the solutions obtained from this special three dimensional case (i.e. axisymmetrical case). In the second case, the perturbation of gravity is equally applied in both the x-z and y-z planes. This means that a true three dimensional convective flow problem is considered in this case. Table 2.1 shows the parameters used in the computations for both cases. To reflect the three-dimensional features of convective pore-fluid flow, the following boundary conditions are used. Temperatures at

Table 2.1 Parameters used for the three-dimensional convective flow problem

Material type	Parameter	Value
pore-fluid	dynamic viscosity	10^{-3} N × s/m^2
	reference density	1000 kg/m^3
	volumetric thermal expansion coefficient	2.07×10^{-4} 1/°C
	specific heat	4185 J/(kg×°C)
	thermal conductivity coefficient	0.6 W/(m×°C)
porous matrix	porosity	0.1
	permeability	10^{-14} m^2
	Specific heat	815 J/(kg×°C)
	thermal conductivity coefficient	3.35 W/(m×°C)

2.6 Application of the Progressive Asymptotic Approach Procedure

the top and the bottom of the computational domain are 0°C and 250°C respectively. Both the top and the bottom of the computational domain are impermeable in the vertical direction, while all the four side boundaries of the computational domain are assumed to be insulated and impermeable in the horizontal direction.

Figures 2.13 and 2.14 show the numerical and analytical solutions for the distributions of pore-fluid velocity and temperature in the axisymmetrical case (i.e. case 1) respectively. As expected, the numerical solutions for both pore-fluid velocity and temperature in this case are exactly axisymmetrical about the y axis. This implies that the three-dimensional problem considered in this particular case can be reasonably treated as a two-dimensional one. It is also observed that the numerical solutions in Fig. 2.13 compare very well with the previous solutions in Fig. 2.14

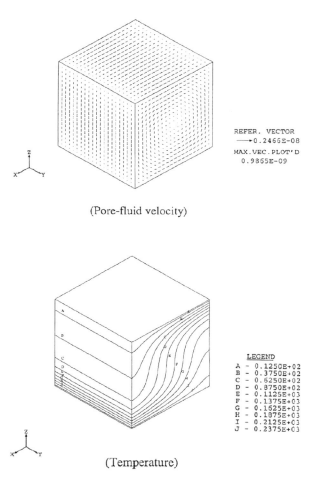

Fig. 2.13 Distributions of pore-fluid velocity and temperature in the porous medium (Case 1, numerical solutions)

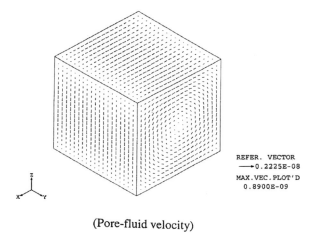

(Pore-fluid velocity)

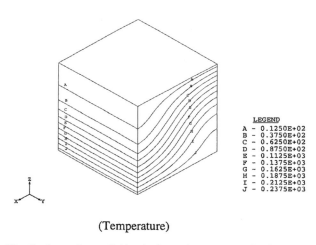

(Temperature)

Fig. 2.14 Distributions of pore-fluid velocity and temperature in the porous medium (Case 1, Analytical solutions)

for the axisymmetrical convective pore-fluid flow problem (Zhao et al. 1997a). This indicates that the progressive asymptotic approach procedure, although it was previously developed for the finite element modelling of two-dimensional convective pore-fluid flow problems, is equally applicable to the finite element modelling of three-dimensional convective pore-fluid flow in fluid-saturated porous media when they are heated from below.

2.6 Application of the Progressive Asymptotic Approach Procedure 33

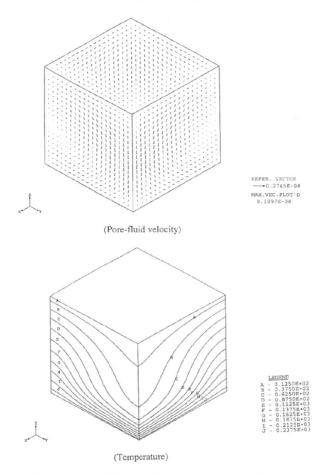

Fig. 2.15 Distributions of pore-fluid velocity and temperature in the porous medium (Case 2)

Figure 2.15 shows the numerical solutions for the distributions of pore-fluid velocity and temperature in the real three-dimensional case (i.e. case 2). By comparing the numerical solutions from case 1 (Fig. 2.13) with those from case 2 (Fig. 2.15), it is observed that the distribution patterns of both pore-fluid velocity and temperature are totally different for these two cases. This demonstrates that the perturbation of gravity at different planes may have a significant effect on the pattern of convective pore-fluid flow in three-dimensional hydrothermal systems. Just like two-dimensional convective flow problems, the solution dependence of three-dimensional convective flow on the direction of the perturbation of gravity need to be considered when these kinds of results are interpreted.

To further observe the movements of pore-fluid particles, the particle tracking technique introduced in this section is used during the numerical computation.

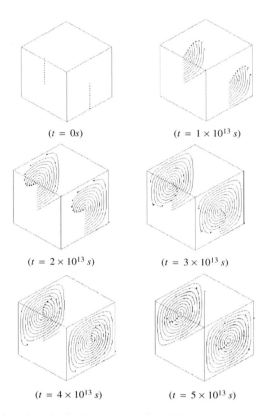

Fig. 2.16 Particle trajectories in the porous medium (Case 1)

We selected 24 particles, 12 of which are in the front plane and the rest are in the back plane of the computation domain, to view the trajectories of those particles. Figure 2.16 shows the particle trajectories in the porous media for several different time instants in case 1. It is clear that in this case, all the particles move within the planes they are initially located within (at $t = 0\,s$), because the convective pore-fluid flow (in case 1) is essentially axisymmetrical about the rotation axis of these particles considered. Figures 2.17 and 2.18 show the particle trajectories in the porous media for several different time instants in case 2. It is observed that the particle trajectories shown in case 2 are totally different from those shown in case 1, although the initial locations (at $t = 0\,s$) of these 24 particles are exactly the same for these two cases. In fact, the particle trajectories shown in case 2 are much more complicated than those shown in case 1. Since pore-fluid is often the sole agent to carry the minerals from the lower crust to the upper crust of the Earth, the different patterns of particle trajectories imply that the pattern of ore body formation and mineralization may be totally different in those two three-dimensional hydrothermal systems.

2.6 Application of the Progressive Asymptotic Approach Procedure

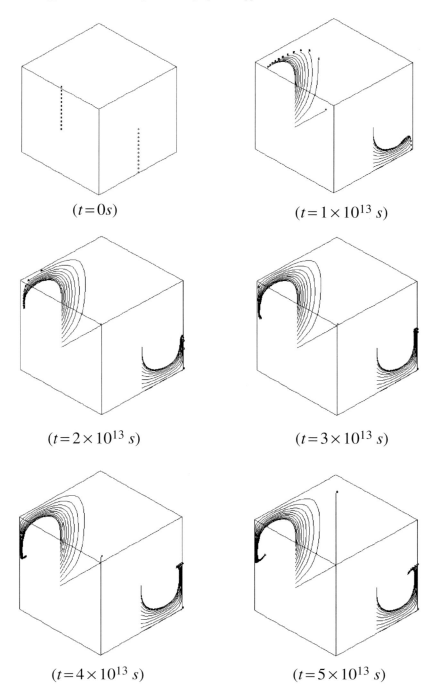

Fig. 2.17 Particle trajectories in the porous medium (Case 2)

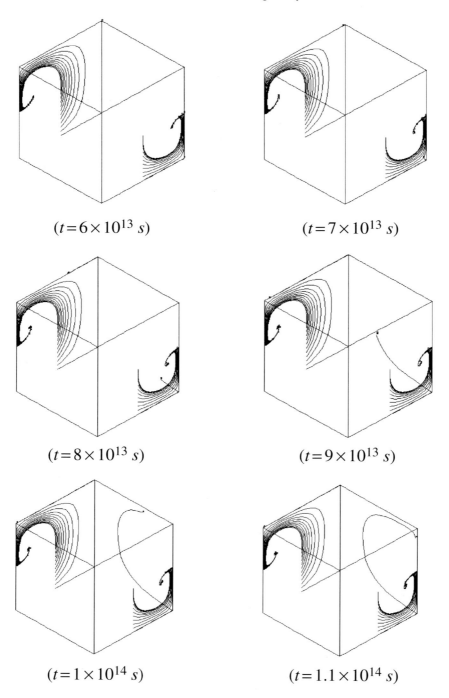

Fig. 2.18 Particle trajectories in the porous medium (Case 2)

Chapter 3
A Consistent Point-Searching Interpolation Algorithm for Simulating Coupled Problems between Deformation, Pore-Fluid Flow, Heat Transfer and Mass Transport Processes in Hydrothemal Systems

Over the past decade or so, many commercial computational codes have become available for solving a great number of practical problems in both scientific and engineering fields. Primary advantages of using commercial computational codes are: (1) built-in pre-processing and post-processing tools make it very easy and attractive to prepare, input and output data which are essential in a numerical analysis; (2) provision of movie/animation functions enables numerical results, the treatment of which is often a cumbersome and tedious task, to be visualised via clear and colourful images; (3) detailed benchmark solutions and documentation as well as many embedded robust solution algorithms allow the codes to be used more easily, correctly, effectively and efficiently for solving a wide range of practical problems. However, the main disadvantage of using commercial computational codes is that each code is often designed, within a certain limit, for solving some particular kinds of practical problems. This disadvantage becomes more and more obvious because the ever-increasing competitiveness in the world economy requires us to deal with more and more complicated and complex geoscience problems, which are encountered and not solved in the field of contemporary computational geoscience. There are three basic ways to overcome the above difficulties. The first is to develop some new commercial computational codes, which is time consuming and often not cost-effective for numerical analysts and consultants. The second is to extend an existing commercial computational code, which is usually impossible because the source code is often not available for the code users. The third is to use several existing commercial computational codes in combination. This requires development of a data translation tool to transfer data necessary between each of the codes to be used. Compared with the difficulties encountered in the first two approaches, the third one is more competitive for most numerical analysts and consultants.

Our first successful example in the practice of using commercial computational codes in a combination manner was to optimize structural topologies under either static or dynamic conditions using the commercial code STRAND6 (G+D Computing 1991) and a home-made code GEMDYN. As a result, a generalized evolutionary method for numerical topological optimization of structures has been developed and many interesting numerical results have been produced (Zhao et al. 1996d, e, 1997b, c, d, 1998c, d). To extend further the idea of using commercial computational codes in combination, we attempt to use a combination of FIDAP (Fluid Dynamics International 1997) and FLAC (Itasca Consulting Group 1995) for solving a fully coupled problem between medium deformation, pore-fluid flow, heat transfer and reactive species transport in a porous medium under high Rayleigh number situations. FIDAP is a well developed, finite element method based, computational fluid dynamics code, whereas FLAC, designed for civil engineering, is based on a finite difference method, but can accommodate unstructured grids. FIDAP can be used to model pore-fluid flow, heat transfer and reactive species transport in a porous medium, but does not treat the medium deformation effects. On the other hand, FLAC is very powerful in its modeling of geomechanical and geological deformation processes, especially for the simulation of large deformation problems. However, the weakness of FLAC is that it cannot be used rigorously to model steady-state pore-fluid convection and the related reactive species transport in a fluid-saturated porous medium. Thus, it is very reasonable to envisage interactively using FIDAP and FLAC for solving a fully coupled problem between medium deformation, pore-fluid flow, heat transfer and reactive mass transport in a porous medium under high Rayleigh number situations. In order to do this, it must be possible to relate accurately any point in the mesh for one code to the equivalent point in the mesh for the other code.

To do this, we present a consistent point-searching interpolation algorithm, also known as the consistent point-searching algorithm for solution interpolation in unstructured meshes consisting of 4-node bilinear quadrilateral elements. The proposed algorithm has the following significant advantages: (1) the use of a point-searching strategy allows a point in one mesh to be accurately related to an element (containing this point) in another mesh. Thus, to translate/transfer the solution of any particular point from mesh 2 to mesh 1, only one element in mesh 2 needs to be inversely mapped. This certainly minimizes the number of elements to which the inverse mapping is applied. In this regard, the proposed consistent algorithm is very effective and efficient. (2) Analytical solutions to the local coordinates of any point in a four-node quadrilateral element, which are derived in a rigorous mathematical manner in the context of this chapter, make it possible to carry out an inverse mapping process very effectively and efficiently. (3) The use of consistent interpolation enables the interpolated solution to be compatible with an original solution and therefore guarantees the interpolated solution of extremely high accuracy. Since the algorithm is very general and robust, it makes it possible to translate and transfer data between FIDAP and FLAC, and vice versa.

3.1 Statement of the Coupled Problem and Solution Method

In terms of simulating the physical and chemical processes associated with ore body formation and mineralization in hydrothermal systems within the upper crust of the Earth, the fully coupled problem between material deformation, pore-fluid flow, heat transfer and mass transport/chemical reactions can be divided into two sub-problems (Zhao et al. 1999c, 2000b). For the first sub-problem, which is the

3.1 Statement of the Coupled Problem and Solution Method

problem describing the steady-state pore-fluid flow, heat transfer and mass transport/chemical reactions in a porous medium, the corresponding governing equations are expressed as follows:

$$\frac{\partial u}{\partial x} + \frac{\partial v}{\partial y} = 0, \tag{3.1}$$

$$u = \frac{K_x}{\mu}\left(-\frac{\partial P}{\partial x}\right), \tag{3.2}$$

$$v = \frac{K_y}{\mu}\left(-\frac{\partial P}{\partial y} + \rho_f g\right), \tag{3.3}$$

$$\rho_{f0} c_p \left(u\frac{\partial T}{\partial x} + v\frac{\partial T}{\partial y}\right) = \lambda_{ex}\frac{\partial^2 T}{\partial x^2} + \lambda_{ey}\frac{\partial^2 T}{\partial y^2}, \tag{3.4}$$

$$\rho_{f0}\left(u\frac{\partial \overline{C}_i}{\partial x} + v\frac{\partial \overline{C}_i}{\partial y}\right) = \rho_{f0}\left(D_{ex}\frac{\partial^2 \overline{C}_i}{\partial x^2} + D_{ey}\frac{\partial^2 \overline{C}_i}{\partial y^2}\right) + \phi R_i \quad (i = 1, 2, \ldots, N), \tag{3.5}$$

$$\rho_f = \rho_{f0}\left[1 - \beta_T(T - T_0) - \sum_{i=1}^{N}\beta_{Ci}(\overline{C}_i - \overline{C}_0)\right], \tag{3.6}$$

$$\lambda_{ex} = \phi\lambda_{fx} + (1-\phi)\lambda_{sx}, \qquad \lambda_{ey} = \phi\lambda_{fy} + (1-\phi)\lambda_{sy}, \tag{3.7}$$

$$D_{ex} = \phi D_{fx}, \qquad D_{ey} = \phi D_{fy}, \tag{3.8}$$

where u and v are the horizontal and vertical velocity components of the pore-fluid in the x and y directions respectively; P is the pore-fluid pressure; T is the temperature of the porous medium; $\overline{C}_i$ is the normalized concentration (in a mass fraction form relative to the pore-fluid density) of chemical species i; N is the total number of the active chemical species considered in the pore-fluid; K_x and K_y are the permeabilities of the porous medium in the x and y directions respectively; μ is the dynamic viscosity of the pore-fluid; ρ_f is the density of the pore-fluid and g is the acceleration due to gravity; ρ_{f0}, T_0 and $\overline{C}_0$ are the reference density, reference temperature and reference normalized concentration of the chemical species used in the analysis; λ_{fx} and λ_{sx} are the thermal conductivities of the pore-fluid and solid matrix in the x direction; λ_{fy} and λ_{sy} are the thermal conductivities of the pore-fluid and solid matrix in the y direction; c_p is the specific heat of the pore-fluid; D_{fx} and D_{fx} are the diffusivities of the chemical species in the x and y directions respectively; ϕ is the porosity of the porous medium; β_T and β_{Ci} are the thermal volume expansion coefficient of the pore-fluid and the volumetric expansion coefficient due to chemical species i; R_i is the chemical reaction term for the transport equation of chemical species i.

Equation (3.6) clearly indicates that the density of the pore-fluid considered in this study is a function of both temperature and chemical species concentrations. This means that the double diffusion effect (Phillips 1991, Nield and Bejan 1992, Alavyoon 1993, Gobin and Bennacer 1994, Nguyen et al. 1994, Goyeau et al. 1996, Nithiarasu et al. 1996, Mamou et al. 1998, Zhao et al. 2000b, 2005b, 2006a) is taken into account in the first sub-problem.

Generally, the chemical reaction term involved in a chemical species transport equation is strongly dependent on the specific chemical reaction considered in the analysis. For a non-equilibrium chemical reaction consisting of aqueous chemical species only, the following type of equation can be considered as an illustrative example.

$$A + B \xrightarrow{k_1} F \qquad (3.9)$$

This equation states that species A and species B react chemically at a rate constant of k_1 and the product of this reaction is species F. For this type of chemical reaction, the reaction term involved in Equation (3.5) can be expressed as follows:

$$R_1 = -k_1 \overline{C}_1 \overline{C}_2, \qquad (3.10)$$

$$R_2 = -k_1 \overline{C}_1 \overline{C}_2, \qquad (3.11)$$

$$R_3 = k_1 \overline{C}_1 \overline{C}_2, \qquad (3.12)$$

where $\overline{C}_1$ and $\overline{C}_2$ are the normalized concentrations (in a mass fraction relative to the pore-fluid density) of species A (i.e. species 1) and B (i.e. species 2) respectively; $\overline{C}_3$ is the normalized concentration (in a mass fraction relative to the pore-fluid density) of species F (i.e. species 3).

Note that for most of the chemical reactions encountered in the field of geoscience, the rate of a chemical reaction is dependent on the temperature at which the chemical reaction takes place. From a geochemical point of view (Nield and Bejan 1992), the temperature dependent nature of the reaction rate for the chemical reaction considered here can be taken into account using the Arrhenius law of the following form:

$$k_1 = k_A \exp\left(\frac{-E_a}{RT}\right) \qquad (3.13)$$

where E_a is the activation energy; R is the gas constant; T is the temperature in Kelvin and k_A is the pre-exponential chemical reaction constant.

The second sub-problem is a static deformation problem under plane strain conditions. If the hydrothermal system is initially in a mechanically equilibrium state, then body forces can be neglected in the corresponding force equilibrium equations. This means that we assume that the material deformation of the hydrothermal system due to gravity has completed before the system is heated by some thermal event from below. Under this assumption, the governing equations for static deformation (which is labeled as the second sub-problem) in the porous medium under plane strain conditions are:

3.1 Statement of the Coupled Problem and Solution Method

$$\frac{\partial \sigma_x}{\partial x} + \frac{\partial \tau_{yx}}{\partial y} = 0, \tag{3.14}$$

$$\frac{\partial \tau_{xy}}{\partial x} + \frac{\partial \sigma_y}{\partial y} = 0, \tag{3.15}$$

$$\sigma_x = \frac{E(1-v)}{(1-2v)(1+v)} \left(\varepsilon_x + \frac{v}{1-v}\varepsilon_y \right) - \frac{E\alpha T}{1-2v} - \tilde{P}, \tag{3.16}$$

$$\sigma_y = \frac{E(1-v)}{(1-2v)(1+v)} \left(\frac{v}{1-v}\varepsilon_x + \varepsilon_y \right) - \frac{E\alpha T}{1-2v} - \tilde{P}, \tag{3.17}$$

$$\tau_{xy} = \tau_{yx} = 2G\gamma_{xy}, \tag{3.18}$$

$$\varepsilon_x = \frac{\partial u_s}{\partial x}, \quad \varepsilon_y = \frac{\partial v_s}{\partial y}, \quad \gamma_{xy} = \frac{1}{2}\left(\frac{\partial u_s}{\partial y} + \frac{\partial v_s}{\partial x}\right), \tag{3.19}$$

where σ_x and σ_y are normal stresses of the solid matrix in the x and y directions; ε_x and ε_y are the normal strains of the solid matrix in relation to σ_x and σ_y; τ_{xy} and γ_{xy} are shear stress and shear strain of the solid matrix; u_s and v_s are the horizontal and vertical displacements of the solid matrix; $\tilde{P}$ is the excess pore-fluid pressure due to the thermal effect; E and G are the elastic and shear modulus respectively; v is the Poisson ratio of the solid matrix and α is the linear thermal expansion coefficient of the solid matrix.

Note that Eqs. (3.14) and (3.15) represent the equilibrium equations, whereas Eqs. (3.16), (3.17), (3.18) and (3.19) are the constitutive equations and strain-displacement relationship equations, respectively.

To couple the first sub-problem with the second sub-problem, we need to establish a relationship between the volumetric strain and the porosity of the porous medium. For small strain problems, such a relationship can be expressed as (Itasca Consulting Group, 1995):

$$\phi = 1 - \frac{1-\phi_0}{1+\varepsilon_v}, \quad \varepsilon_v = \varepsilon_x + \varepsilon_y, \tag{3.20}$$

where ϕ and ϕ_0 are the porosity and initial porosity of the porous medium; ε_v is the volumetric strain of the solid matrix.

Using the Carman-Kozeny formula (Nield and Bejan 1992), the permeability of an isotropic porous medium is expressed as a function of porosity as follows:

$$K_x = K_y = \frac{K_0(1-\phi_0)^2\phi^3}{\phi_0^3(1-\phi)^2}, \tag{3.21}$$

where K_0 is the initial permeability corresponding to the initial porosity, ϕ_0.

Obviously, the first sub-problem is coupled with the second sub-problem through the medium temperature, T, pore-fluid pressure, p, and permeabilities, K_x and K_y.

Within the first sub-problem itself, the pore-fluid flow is coupled with the thermal flow and reactive flow through the medium temperature, aqueous chemical species concentrations and pore-fluid velocity components, u and v. However, within the second sub-problem itself, the medium deformation (displacement) is coupled with the medium temperature, pore-fluid pressure and permeability through the volumetric strain, ε_v, and the medium porosity, ϕ.

With regard to the general solution method used in this study, FIDAP (Fluid Dynamics International 1997), which is one of the most powerful finite element codes currently available for dealing with a very wide range of practical problems in fluid dynamics, is used to solve the first sub-problem, while FLAC (Itasca Consulting Group 1995), which is a very useful finite difference code for dealing with material deformation problems in the fields of both geotechnical engineering and geoscience, is used to solve the second sub-problem. A general interface program is developed to translate and transfer data between FIDAP and FLAC. To obtain non-trivial (convective) solutions for the pore-fluid flow under supercritical Rayleigh number conditions, the progressive asymptotic approach procedure discussed in Chap. 2 (Zhao et al. 1997a) is also employed in the computation.

3.2 Mathematical Formulation of the Consistent Point-Searching Interpolation Algorithm in Unstructured Meshes

The motivation of developing a consistent point-searching interpolation algorithm is described as follows. For two totally different meshes, namely mesh 1 (mesh in FIDAP) and mesh 2 (mesh in FLAC) shown in Fig. 3.1, finite element solutions to the temperature and pressure due to hydrothermal effects are available (from FIDAP) at the nodal points of mesh 1 but not known at those of mesh 2. This

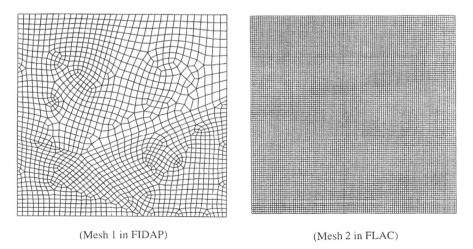

(Mesh 1 in FIDAP)   (Mesh 2 in FLAC)

Fig. 3.1 Two totally different meshes used in FIDAP and FLAC

3.2 Mathematical Formulation of the Algorithm

indicates that to evaluate the thermal deformation using mesh 2 in FLAC, it is essential to translate/transfer the nodal solutions from mesh 1 into mesh 2. Since mesh 1 is totally different from mesh 2, it is only possible to interpolate the nodal solution of mesh 2, point by point, using the nodal solutions of mesh 1 and the related elemental information. Thus, for any nodal point in mesh 2, it is necessary to find out the element, which contains this particular point, in mesh 1. This is the first step, termed point-searching, in the proposed algorithm. Once a point in mesh 2 is related to an element (containing this point) in mesh 1, the local coordinates of this point need to be evaluated using the related information of the element in mesh 1. This requires an inverse mapping to be carried out for this particular element in mesh 1, so that the nodal solution of the point in mesh 2 can be consistently interpolated in the finite element sense. This is the main reason why the proposed algorithm is called the consistent point-searching interpolation algorithm.

Clearly, for the purpose of developing the consistent point-searching interpolation algorithm, one has to deal with the following two key issues. First, an efficient searching strategy needs to be developed to limit the number of elements, to which the inverse mapping is applied, so that the algorithm growth rate can be reduced to the minimum. Second, the issue of the parametric inverse mapping between the real (global) and element (local) coordinates should be dealt with in an appropriate manner. Generally, the inverse mapping problem is a nonlinear one, which requires numerical solutions for higher order elements. However, for 4-node bilinear quadrilateral elements, it is possible to solve this problem analytically. Since most practical problems in finite element analysis can be modelled reasonably well using 4-node bilinear quadrilateral elements, the analytical solution to the inverse mapping problem for this kind of element may find wide applications for many practical problems. Given the importance of the above two key issues, they are addressed in great detail in the following sections, respectively.

3.2.1 Point Searching Step

In this section, we deal with the first issue, i.e., the development of an efficient searching strategy to limit the number of elements, which need to be inversely mapped. To achieve this, we have to establish a strategy to relate a point in one mesh to an element in another mesh accurately. Consider a four-node quadrilateral element shown in Fig. 3.2. If any point (i.e., point A in Fig. 3.2) is located within this element, then the following equations hold true:

$$\vec{12} \times \vec{1A} = \lambda_1 \vec{k}, \tag{3.22}$$

$$\vec{23} \times \vec{2A} = \lambda_2 \vec{k}, \tag{3.23}$$

$$\vec{34} \times \vec{3A} = \lambda_3 \vec{k}, \tag{3.24}$$

$$\vec{41} \times \vec{4A} = \lambda_4 \vec{k}, \tag{3.25}$$

$$\lambda_i \geq 0 \qquad (i = 1, 2, 3, 4), \tag{3.26}$$

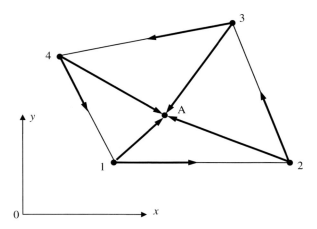

Fig. 3.2 Point A in a fournode quadrilateral element

where $\vec{12}$, $\vec{23}$, $\vec{34}$ and $\vec{41}$ are vectors of four sides of the element; $\vec{1A}$, $\vec{2A}$, $\vec{3A}$ and $\vec{4A}$ are vectors of each node of the element to point A respectively; $\vec{k}$ is a normal vector of the plane where the element is located; λ_i ($i = 1, 2, 3, 4$) are four different constants. Note that the left-hand sides of Eqs. (3.22), (3.23), (3.24) and (3.25) represent the cross products of two vectors.

By using the global coordinates of point A and four nodal points of the element, Eqs (3.22), (3.23), (3.24), (3.25) and (3.26) can be expressed in the following form:

$$(x_2 - x_1)(y_A - y_1) - (x_A - x_1)(y_2 - y_1) = \lambda_1 \geq 0, \quad (3.27)$$

$$(x_3 - x_2)(y_A - y_2) - (x_A - x_2)(y_3 - y_2) = \lambda_2 \geq 0, \quad (3.28)$$

$$(x_4 - x_3)(y_A - y_3) - (x_A - x_3)(y_4 - y_3) = \lambda_3 \geq 0, \quad (3.29)$$

$$(x_1 - x_4)(y_A - y_4) - (x_A - x_4)(y_1 - y_4) = \lambda_4 \geq 0, \quad (3.30)$$

where x_i and y_i are the global coordinates of nodal points of the element; x_A and y_A are the global coordinates of point A.

Note that when point A is located on the side of the element, two vectors related to a node of the element are coincident. Consequently, the corresponding λ_i to this node must be equal to zero. Since Eqs. (3.27), (3.28), (3.29) and (3.30) are only dependent on the global coordinates of five known points (i.e., point A and four nodes of the element), they can be straightforwardly used to predict the element, in which point A is located.

It must be pointed out that, using the point-searching strategy expressed by Eqs. (3.27), (3.28), (3.29) and (3.30), it is very easy and accurate to relate a point in mesh 2 to an element (containing this point) in mesh 1, since Eqs. (3.27), (3.28), (3.29) and (3.30) only involve certain simple algebraic calculations which can be carried out by computers at a very fast speed. Thus, to translate/transfer the solution

3.2 Mathematical Formulation of the Algorithm

of any point from mesh 1 to mesh 2, only one element in mesh 1 needs to be inversely mapped. This certainly reduces the number of elements, to which the inverse mapping is applied, to the minimum. From the computational point of view, the present algorithm is both effective and efficient.

3.2.2 Inverse Mapping Step

This section deals with the second key issue related to the development of the consistent point-searching interpolation algorithm. In order to interpolate the solution at point A consistently in the finite element sense, it is necessary to find out the local coordinates of this point by an inverse mapping. For a four-node quadrilateral isoparametric element shown in Fig. 3.3, it is possible to implement the inverse mapping analytically. The following gives the related mathematical equations.

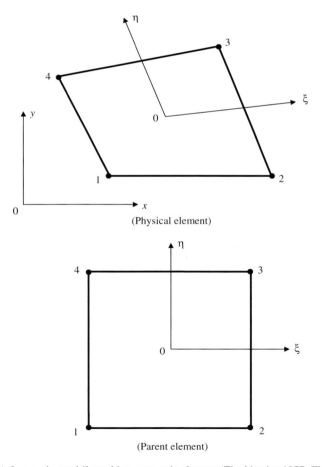

Fig. 3.3 A four-node quadrilateral isoparametric element (Zienkiewicz 1977; Zhao et al. 1999f)

The forward mapping of the physical element in the global system to the parent element in the local system reads:

$$x = \sum_{i=1}^{4} N_i x_i, \tag{3.31}$$

$$y = \sum_{i=1}^{4} N_i y_i, \tag{3.32}$$

where

$$N_1 = \frac{1}{4}(1 - \xi)(1 - \eta), \tag{3.33}$$

$$N_2 = \frac{1}{4}(1 + \xi)(1 - \eta), \tag{3.34}$$

$$N_3 = \frac{1}{4}(1 + \xi)(1 + \eta), \tag{3.35}$$

$$N_4 = \frac{1}{4}(1 - \xi)(1 + \eta), \tag{3.36}$$

where N_1, N_2, N_3 and N_4 are shape functions of the element.

For a given point (i.e., point A), the corresponding local coordinates are ξ_A and η_A. Substituting these local coordinates into Eqs. (3.31), (3.32), (3.33), (3.34), (3.35) and (3.36) yields the following inverse mapping:

$$a_2 \xi_A + a_3 \eta_A + a_4 \xi_A \eta_A = 4x_A - a_1, \tag{3.37}$$

$$b_2 \xi_A + b_3 \eta_A + b_4 \xi_A \eta_A = 4y_A - b_1, \tag{3.38}$$

where

$$a_1 = x_1 + x_2 + x_3 + x_4, \tag{3.39}$$

$$a_2 = -x_1 + x_2 + x_3 - x_4, \tag{3.40}$$

$$a_3 = -x_1 - x_2 + x_3 + x_4, \tag{3.41}$$

$$a_4 = x_1 - x_2 + x_3 - x_4, \tag{3.42}$$

$$b_1 = y_1 + y_2 + y_3 + y_4, \tag{3.43}$$

$$b_2 = -y_1 + y_2 + y_3 - y_4, \tag{3.44}$$

$$b_3 = -y_1 - y_2 + y_3 + y_4, \tag{3.45}$$

$$b_4 = y_1 - y_2 + y_3 - y_4. \tag{3.46}$$

3.2 Mathematical Formulation of the Algorithm

Since Eqs. (3.37) and (3.38) are two second-order simultaneous equations, the solution to the local coordinates, ξ_A and η_A, can be obtained in the following different cases.

3.2.2.1 Case 1: $a_4 = 0, b_4 = 0$

In this case, Eqs. (3.37) and (3.38) can be written as

$$a_2\xi_A + a_3\eta_A = c_1, \tag{3.47}$$

$$b_2\xi_A + b_3\eta_A = c_2, \tag{3.48}$$

where

$$c_1 = 4x_A - a_1, \tag{3.49}$$

$$c_2 = 4y_A - b_1. \tag{3.50}$$

Clearly, Eqs. (3.47) and (3.48) are a set of standard linear simultaneous equations, so that the corresponding solutions can be immediately obtained as

$$\xi_A = \frac{b_3 c_1 - a_3 c_2}{a_2 b_3 - a_3 b_2}, \tag{3.51}$$

$$\eta_A = \frac{-b_2 c_1 + a_2 c_2}{a_2 b_3 - a_3 b_2}. \tag{3.52}$$

3.2.2.2 Case 2: $a_4 = 0, b_4 \neq 0$

The corresponding equations in this case can be rewritten as

$$a_2\xi_A + a_3\eta_A = c_1, \tag{3.53}$$

$$b_2\xi_A + b_3\eta_A + b_4\xi_A\eta_A = c_2. \tag{3.54}$$

The solutions to Eqs. (3.53) and (3.54) can be expressed for the following three sub-cases:

(1) $a_2 = 0, a_3 \neq 0$

$$\eta_A = \frac{c_1}{a_3}, \tag{3.55}$$

$$\xi_A = \frac{c_2 - b_3\eta_A}{b_2 + b_4\eta_A}. \tag{3.56}$$

(2) $a_2 \neq 0, a_3 = 0$

$$\xi_A = \frac{c_1}{a_2}, \tag{3.57}$$

$$\eta_A = \frac{c_2 - b_2 \xi_A}{b_3 + b_4 \xi_A}. \tag{3.58}$$

(3) $a_2 \neq 0,\ a_3 \neq 0$

$$\eta_A = \frac{-bb \pm \sqrt{bb^2 - (4aa)cc}}{2aa}, \tag{3.59}$$

$$\xi_A = \frac{c_1 - a_3 \eta_A}{a_2}, \tag{3.60}$$

where

$$aa = \frac{b_4 a_3}{a_2}, \tag{3.61}$$

$$bb = \frac{b_2 a_3 - b_4 c_1}{a_2} - b_3, \tag{3.62}$$

$$cc = \frac{-b_2 c_1}{a_2}. \tag{3.63}$$

3.2.2.3 Case 3: $a_4 \neq 0,\ b_4 = 0$

This case is very similar to case 2. The corresponding equations in this case are as follows:

$$a_2 \xi_A + a_3 \eta_A + a_4 \xi_A \eta_A = c_1, \tag{3.64}$$

$$b_2 \xi_A + b_3 \eta_A = c_2. \tag{3.65}$$

Similarly, the solution to Equations (3.64) and (3.65) can be expressed in the following three sub-cases:

(1) $b_2 = 0,\ b_3 \neq 0$

$$\eta_A = \frac{c_2}{b_3}, \tag{3.66}$$

$$\xi_A = \frac{c_1 - a_3 \eta_A}{a_2 + a_4 \eta_A}. \tag{3.67}$$

(2) $b_2 \neq 0,\ b_3 = 0$

$$\xi_A = \frac{c_2}{b_2}, \tag{3.68}$$

$$\eta_A = \frac{c_1 - a_2 \xi_A}{a_3 + a_4 \xi_A}. \tag{3.69}$$

3.2 Mathematical Formulation of the Algorithm

(3) $b_2 \neq 0,\ b_3 \neq 0$

$$\eta_A = \frac{-bb \pm \sqrt{bb^2 - (4aa)cc}}{2aa}, \tag{3.70}$$

$$\xi_A = \frac{c_2 - b_3\eta_A}{b_2}, \tag{3.71}$$

where

$$aa = \frac{a_4 b_3}{b_2}, \tag{3.72}$$

$$bb = \frac{a_2 b_3 - a_4 c_2}{b_2} - b_3, \tag{3.73}$$

$$cc = \frac{-a_2 c_2}{b_2}. \tag{3.74}$$

3.2.2.4 Case 4: $a_4 \neq 0,\ b_4 \neq 0$

In this case, Eqs. (3.37) and (3.38) can be written in the following form:

$$a_2^* \xi_A + a_3^* \eta_A + \xi_A \eta_A = c_1^*, \tag{3.75}$$

$$b_2^* \xi_A + b_3^* \eta_A + \xi_A \eta_A = c_2^*, \tag{3.76}$$

where

$$a_i^* = \frac{a_i}{a_4}, \quad b_i^* = \frac{b_i}{b_4} \quad (i = 2, 3), \tag{3.77}$$

$$c_1^* = \frac{c_1}{a_4}, \quad c_2^* = \frac{c_2}{b_4}. \tag{3.78}$$

Subtracting Eq. (3.78) from Eq. (3.77) yields the following equation:

$$(a_2^* - b_2^*)\xi_A + (a_3^* - b_3^*)\eta_A = c_1^* - c_2^*. \tag{3.79}$$

Clearly, if $a_2^* - b_2^* = 0$, the solutions to Eqs. (3.77) and (3.78) can be straightforwardly expressed as

$$\eta_A = \frac{c_1^* - c_2^*}{a_3^* - b_3^*}, \tag{3.80}$$

$$\xi_A = \frac{c_1^* - a_3^* \eta_A}{a_2^* + \eta_A}. \tag{3.81}$$

Otherwise, Eq. (3.79) can be rewritten as

$$\xi_A = \alpha - \beta \eta_A \tag{3.82}$$

where

$$\alpha = \frac{c_1^* - c_2^*}{a_2^* - b_2^*}, \tag{3.83}$$

$$\beta = \frac{a_3^* - b_3^*}{a_2^* - b_2^*}. \tag{3.84}$$

Therefore, the corresponding solutions can be expressed for the following subcases.

(1) For $\beta = 0$

$$\xi_A = \alpha, \tag{3.85}$$

$$\eta_A = \frac{c_1^* - a_2^* \xi_A}{a_3^* + \xi_A}. \tag{3.86}$$

(2) For $\beta \neq 0$

$$\eta_A = \frac{-(a_2^* \beta - a_3^* - \alpha) \pm \sqrt{(a_2^* \beta - a_3^* - \alpha)^2 - 4\beta(c_1^* - a_2^* \alpha)}}{2\beta}, \tag{3.87}$$

$$\xi_A = \alpha - \beta \eta_A. \tag{3.88}$$

3.2.3 Consistent Interpolation Step

Based on the concept of isoparametric elements (Zienkiewicz 1977), any nodal solution at any point (i.e., point A) in mesh 1 can be consistently interpolated using the following equation:

$$S_A = \sum_{i=1}^{4} N_i(\xi_A, \eta_A) S_i, \tag{3.89}$$

where S_i is the appropriate numerical solution at a nodal point in mesh 1.

Note that since the global coordinate system used in mesh 1 is exactly the same as in mesh 2, the nodal value of any concerned solution in mesh 2 can be interpolated through mesh 1 and then directly transferred to mesh 2.

3.3 Verification of the Proposed Consistent Point-Searching Interpolation Algorithm

In order to verify the proposed consistent point-searching interpolation algorithm in two totally different meshes, we consider a coupled problem between medium deformation, pore-fluid flow and heat transfer processes in a fluid-saturated porous elastic medium. The square computational domain is 10×10 km^2 in size. This computational domain is discretized into mesh 1 of 1625 nodes and mesh 2 of 10000 nodes respectively. Firstly, we use FIDAP with mesh 1 to simulate the high Rayleigh number convection cells, temperature and pressure distributions in the computational domain. Then, we use the proposed algorithm to translate/transfer the temperature and pressure solutions in mesh 1 into FLAC's mesh, mesh 2.

To simulate the high Rayleigh number convection cells in the computational domain of the test problem, the following parameters are used in the computation. For pore-fluid, dynamic viscosity is 10^{-3} N $\times$ s/m^2; reference density is 1000 kg/m^3; volumetric thermal expansion coefficient is $2.07 \times 10^{-4}(1/^0C)$; specific heat is 4185 J/(kg$\times ^0$C); thermal conductivity coefficient is 0.6 W/(m$\times ^0$C) in both the horizontal and vertical directions. For the porous matrix, initial porosity is 0.1; initial permeability is 10^{-14}m^2; thermal conductivity coefficient is 3.35 W/(m$\times ^0$C) in both the horizontal and vertical directions. The temperature is 25°C at the top of the domain, while it is 225°C at the bottom of the domain. This means that the computational domain is heated uniformly from below. The left and right lateral boundaries of the computational domain are insulated and impermeable in the horizontal direction, whereas the top and bottom are impermeable in the vertical direction.

Figures 3.4 and 3.5 show the comparison of the original temperature and pressure solutions in mesh 1 with the corresponding translated/ transferred solutions in mesh 2. It is clear that although mesh 2 is totally different from mesh 1, the translated/transferred solutions in mesh 2 are essentially the same as in mesh 1. This demonstrates the correctness and effectiveness of the proposed consistent point-searching interpolation algorithm.

To further test the robustness of the proposed algorithm, we use the concept of a transform in mathematics below. For instance, the robustness of a numerical Fourier transform algorithm is often tested by the following procedure: (1) implementation of a forward Fourier transform to an original function; and (2) implementation of an inverse Fourier transform to the forwardly transformed function. If the inversely transformed function is exactly the same as the original function, it demonstrates that the numerical Fourier transform algorithm is robust. Clearly, the same procedure as above can be followed to examine the robustness of the proposed consistent point-searching algorithm. For this purpose, the forward transform is defined as translating/transferring solution data from mesh 1 to mesh 2, while the inverse transform is defined as translating/transferring the translated/transferred solution data back from mesh 2 to mesh 1.

Figures 3.6 and 3.7 show the comparisons of the forwardly transformed temperature and pressure solutions (from mesh 1 to mesh 2) with the inversely transformed

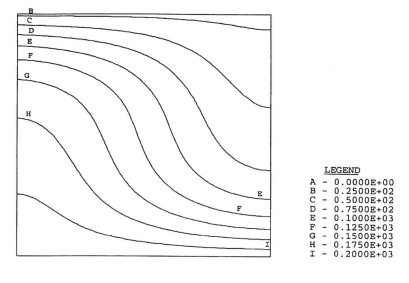

(Original solution in mesh 1)

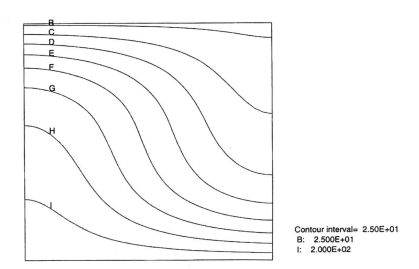

(Translated/transferred solution in mesh 2)

Fig. 3.4 Comparison of the original solution with the translated/transferred solution (Temperature)

3.3 Verification of the Proposed Consistent Point-Searching Interpolation Algorithm

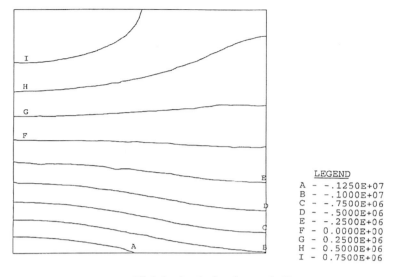

LEGEND
A - -.1250E+07
B - -.1000E+07
C - -.7500E+06
D - -.5000E+06
E - -.2500E+06
F - 0.0000E+00
G - 0.2500E+06
H - 0.5000E+06
I - 0.7500E+06

(Original solution in mesh 1)

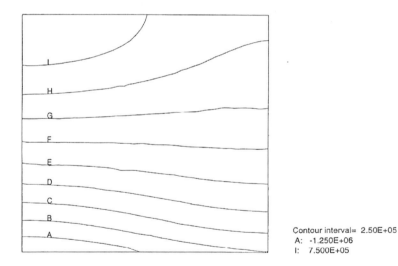

Contour interval= 2.50E+05
A: -1.250E+06
I: 7.500E+05

(Translated/transferred solution in mesh 2)

Fig. 3.5 Comparison of the original solution with the translated/transferred solution (Pore-fluid pressure)

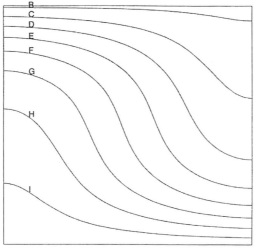

(Forwardly transformed solution)

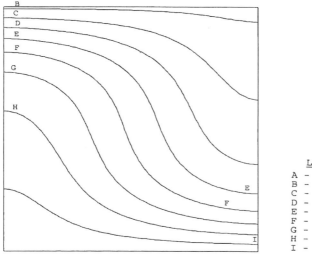

(Inversely transformed solution)

Fig. 3.6 Comparison of the forwardly transformed solution with the inversely transformed solution (Temperature)

3.3 Verification of the Proposed Consistent Point-Searching Interpolation Algorithm   55

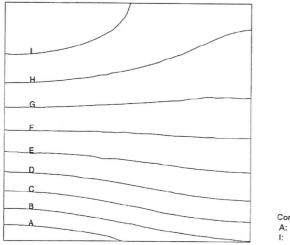

(Forwardly transformed solution)

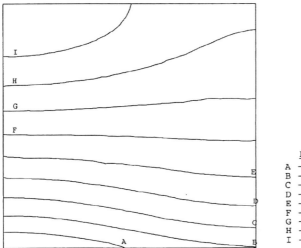

(Inversely transformed solution)

Fig. 3.7 Comparison of the forwardly transformed solution with the inversely transformed solution (Pore-fluid pressure)

solutions (from mesh 2 to mesh 1). Clearly, the forwardly transformed solutions in mesh 2 are exactly the same as the inversely transformed solutions in mesh 1. Furthermore, the inversely transformed solutions in mesh 1 (see Figs. 3.6 and 3.7) also compare very well with the original solutions in mesh 1 (see Figs. 3.4 and 3.5). This means that after the original solutions are transformed from mesh 1 to mesh 2, they can be transformed back exactly from mesh 2 to mesh 1. Such a reversible process demonstrates the robustness of the proposed consistent point-searching interpolation algorithm.

3.4 Application Examples of the Proposed Consistent Point-Searching Interpolation Algorithm

3.4.1 Numerical Modelling of Coupled Problems Involving Deformation, Pore-Fluid Flow and Heat Transfer in Fluid-Saturated Porous Media

Since the verification example considered in Sect. 3.3 is a coupled problem between medium deformation, pore-fluid flow and heat transfer processes in a fluid-saturated porous elastic medium, it can be used as the first application example of the proposed consistent point-searching interpolation algorithm. Thus, we can continue the simulation of the verification example and use FLAC with the translated/transferred temperature in mesh 2 to compute thermal deformation and stresses in the fluid-saturated porous elastic medium. Towards this end, it is assumed that: (1) the bottom boundary of the computational domain is fixed; (2) the top boundary is free; and (3) the two lateral boundaries are horizontally fixed but vertically free. Except for the parameters used in Sect. 3.3, the following additional parameters are used in the continued simulation: the elastic modulus of the porous medium is 1×10^{10} Pa; Poisson's ratio is 0.25; the volumetric thermal expansion coefficient is $2.07 \times 10^{-4} (1/^{\circ}C)$ and the initial porosity is 0.1.

Figures 3.8 and 3.9 show the temperature induced deformation and stresses in the porous elastic medium respectively. It is observed that the distribution pattern of the volumetric strain is similar to that of the temperature. That is to say, higher temperature results in larger volumetric strain, as expected from the physics point of view. Owing to relatively larger volumetric strain in the left side of the computational domain, the maximum vertical displacement takes place at the upper left corner of the domain. This is clearly evidenced in Fig. 3.8. As a direct consequence of the thermal deformation, the temperature induced horizontal stress dominates in the porous medium. The maximum horizontal compressive stress due to the thermal effect takes place in the hottest region of the computational domain, while the maximum horizontal tensile stress occurs at the upper left corner of the domain, where the vertical displacement reaches its maximum value. Apart from a small part of the top region of the computational domain, the thermal induced vertical stress is

3.4 Examples of the Proposed Consistent Point-Searching Interpolation Algorithm   57

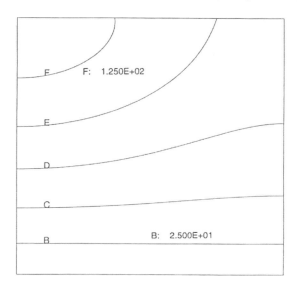

(Vertical displacement, Contour interval = 2.5m)

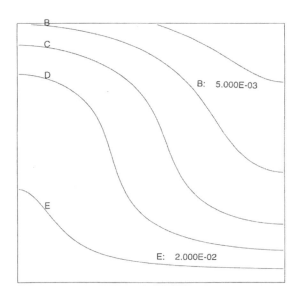

(Volumetric strain, Contour interval = 5.0e-03)

Fig. 3.8 Deformation of the porous medium due to temperature effect

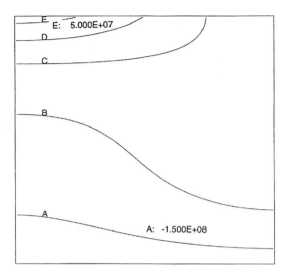

(Horizontal stress, Contour interval = 5.0e+07 Pa)

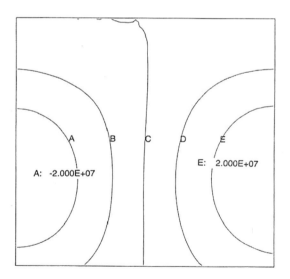

(Vertical stress, Contour interval = 1.0e+07 Pa)

Fig. 3.9 Temperature induced stresses in the porous medium

well distributed anti-symmetrically. This indicates that the vertical forces are well balanced in the horizontal cross-section of the computational domain.

3.4.2 Numerical Modelling of Coupled Problems Involving Deformation, Pore-Fluid Flow, Heat Transfer and Mass Transport in Fluid-Saturated Porous Media

Generally, the numerical algorithms and methods can be used to answer many what-if questions related to ore body formation and mineralization that can be described as coupled problems between medium deformation, pore-fluid flow, heat transfer and reactive mass transport in fluid-saturated porous media. However, due to the approximate nature of a numerical solution, it is essential to evaluate the accuracy of the numerical solution, at least qualitatively if the analytical solution to the problem is not available. This requires us to have a strong theoretical understanding of the basic governing principles and processes behind the coupled problem. Specifically, we must know, through some kind of theoretical analysis, what the pore-fluid can and cannot do, and what reaction patterns the reactive pore-fluid can produce in a hydrothermal system. For this reason, we have been making efforts, in recent years, to develop theoretical solutions to verify the numerical methods developed (Zhao and Valliappan 1993a, b, 1994a, b, Zhao et al. 1997a, 1998a, 1999b). On the other hand, a good numerical solution can provide insights into the integrated behaviour of different processes that occur in a hydrothermal system. Even in some circumstances, a good numerical solution can provide some useful hints for deriving analytical solutions to some aspects of the problem. This indicates that the numerical and theoretical approaches are, indeed, complementary in the field of computational geoscience. Realizing this particular relationship between the analytical method and the numerical method, Phillips (1991) stated that: "A conceptual framework elucidating the relations among flow characteristics, driving forces, structure, and reaction patterns enables us not only to understand the results of numerical modelling more clearly, but to check them. (Numerical calculations can converge to a grid-dependent limit, and artifacts of a solution can be numerical rather than geological.) Numerical modelling provides a quantitative description and synthesis of a basin-wide flow in far greater detail than would be feasible analytically. The combination of the two techniques is a much more powerful research tool than either alone." Keeping this in mind, we have used the proposed consistent point-searching interpolation algorithm to develop a general interface between the two commercial computational codes, FIDAP and FLAC. This development enables us to investigate the integrated behaviour of ore body formation and mineralization in hydrothermal systems.

The first application example, which is closely associated with the coupled problem between medium deformation, pore-fluid flow, heat transfer and reactive mass transport in a fluid-saturated porous medium, is to answer the question: What is the pattern of pore-fluid flow, the distributions of temperature, reactant and product

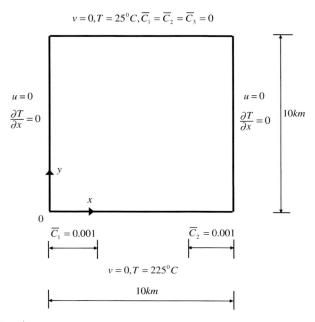

Fig. 3.10 Geometry and boundary conditions of the coupled problem

chemical species, and the patterns of final porosity and permeability, if a deformable hydrothermal system has two reservoirs for two different reactant chemical species, and is heated uniformly from below?

As shown in Fig. 3.10, the computational domain considered for this example is a square box of 10 by 10 km in size. The temperature at the top of the domain is 25°C, while it is 225°C at the bottom of the domain. The left and right lateral boundaries are insulated and impermeable in the horizontal direction, whereas the top and bottom are impermeable in the vertical direction. To consider the mixing of reactant chemical species, the normalized concentration of species 1 is 0.001 at the left quarter of the bottom and the normalized concentration of species 2 is 0.001 at the right quarter of the bottom. This implies that species 1 and 2 are injected into the computational domain through two different reservoirs at different locations. The normalized concentrations of all three species involved in the chemical reaction are assumed to be zero at the top surface of the computational domain. Table 3.1 shows the related parameters used in the computation. In addition, the computational domain is discretized into 2704 quadrilateral elements with 2809 nodes in total.

Figure 3.11 shows the pore-fluid velocity and temperature distributions in the deformable porous medium. It is observed that a clockwise convection cell has formed in the pore-fluid flow in the porous medium. This convective flow results in the localized distribution of temperature, as can be seen clearly from Fig. 3.11. As we mentioned in the beginning of this section, we have to validate the numerical solution, at least from a qualitative point of view. For the hydrothermal system

3.4 Examples of the Proposed Consistent Point-Searching Interpolation Algorithm

Table 3.1 Parameters used for the first coupled problem involving reactive mass transport

Material type	Parameter	Value
pore-fluid	dynamic viscosity	$10^{-3}\,\text{N} \times \text{s/m}^2$
	reference density	$1000\,\text{kg/m}^3$
	volumetric thermal expansion coefficient	$2.07 \times 10^{-4}(1/^0\text{C})$
	specific heat	$4185\,\text{J/(kg}\times{^0}\text{C})$
	thermal conductivity coefficient	$0.6\,\text{W/(m}\times{^0}\text{C})$
	chemical species diffusivity coefficient	$3 \times 10^{-6}\,\text{m}^2/\text{s}$
	pre-exponential reaction rate constant	$10^{-7}\,\text{kg/(m}^3\times\text{s})$
	activate energy of reaction	$5 \times 10^4\,\text{J/mol}$
	gas constant	$8.315\,\text{J/(mol}\times{^0}\text{K})$
porous matrix	initial porosity	0.1
	initial permeability	$10^{-14}\,\text{m}^2$
	elastic modulus	$2 \times 10^{10}\,\text{Pa}$
	Poisson's ratio	0.25
	volumetric thermal expansion coefficient	$2.07 \times 10^{-5}(1/^0\text{C})$
	thermal conductivity coefficient	$3.35\,\text{W/(m}\times{^0}\text{C})$

studied, the temperature gradient is the main driving force to initiate the convective pore-fluid flow. There is a criterion available (Phillips 1991, Nield and Bejan 1992, Zhao et al. 1997a) for judging whether or not the convective pore-fluid flow is possible in such a hydrothermal system as considered here. The criterion says that if the Rayleigh number of the hydrothermal system, which has a flat bottom and is heated uniformly from below, is greater than the critical Rayleigh number, which has a theoretical value of $4\pi^2$, then the convective pore-fluid flow should take place in the hydrothermal system considered. For the particular hydrothermal system considered in this application example, the Rayleigh number is 55.2, which is greater than the critical Rayleigh number. Therefore, the convective pore-fluid flow should occur in the hydrothermal system, from the theoretical point of view. This indicates that the numerical solution obtained from this application example can be qualitatively justified.

Figure 3.12 shows the normalized concentration distributions of reactant and product chemical species, whereas Fig. 3.13 shows the final distribution of porosity and permeability in the computational domain of this application example, where both the material deformation due to thermal effects and the reactive mass transport are taken into account in the numerical computation. Obviously, the distributions of the three chemical species are different. This indicates that the convective pore-fluid flow may significantly affect the chemical reactions (i.e. reaction flow) in the deformable porous medium. In particular, the distribution of chemical species 3, which is the product of the chemical reaction considered in this application example, demonstrates that the produced chemical species in the chemical reaction may reach its equilibrium concentration in some regions, but may not necessarily reach its equilibrium concentration in other regions in the computational domain. This finding might be important for the further understanding of reactive flow transport

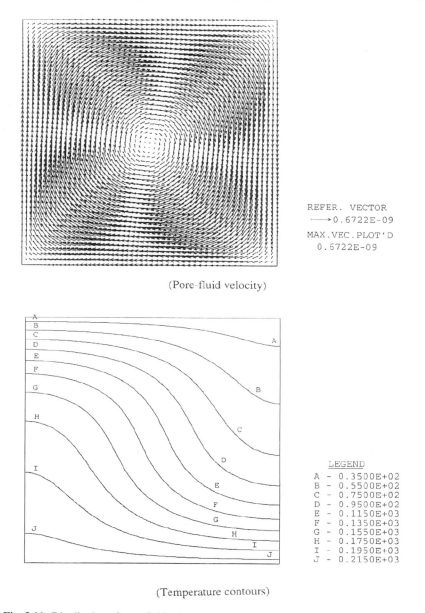

Fig. 3.11 Distribution of pore-fluid velocity and temperature in the hydrothermal system

3.4 Examples of the Proposed Consistent Point-Searching Interpolation Algorithm 63

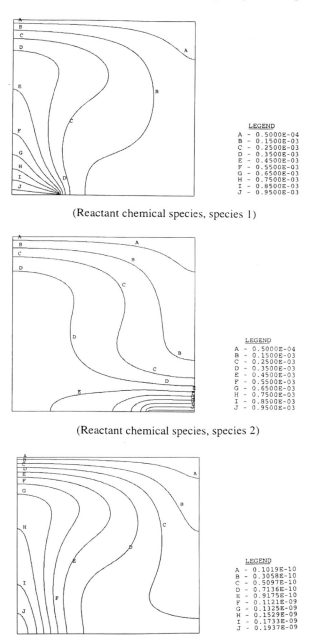

Fig. 3.12 Normalized concentration distribution of reactant and production chemical species in the hydrothermal system

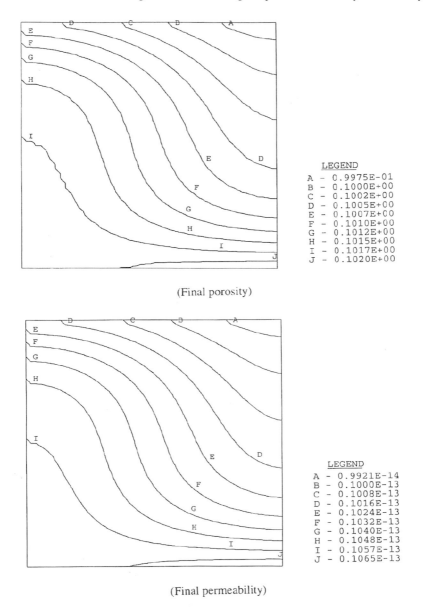

(Final porosity)

(Final permeability)

Fig. 3.13 Distribution of final porosity and permeability in the hydrothermal system

3.4 Examples of the Proposed Consistent Point-Searching Interpolation Algorithm 65

involved in ore body formation and mineralization in the upper crust of the Earth. Since the feedback effect of the medium deformation on the porosity and permeability has been considered in the numerical simulation, the distributions of the final porosity and permeability are totally different from that of their initial values, which are uniformly distributed in the computational domain.

Again, we can justify the numerical solution for the final porosity and permeability in the hydrothermal system. From the physics point of view, we know that the higher the relative temperature in a porous medium, the larger the deformation of the porous medium due to the thermal effect. On the other hand, since the solid aggregates in the porous medium are relatively stiff, the larger the deformation of the porous medium, the greater the porosity of the porous medium, indicating that large deformation of a porous medium results in greater permeability of the porous medium. This implies that for a porous medium, a region of relatively high temperature favours the formation of flow channels because of an increase in the porosity of the region due to the thermal effect. This kind of phenomenon can be clearly observed from the related numerical solutions for the temperature distribution (Fig. 3.11) and the final porosity/permeability distributions (Fig. 3.13) in the hydrothermal system. This further justifies the numerical solutions obtained from this application example, at least from a qualitative point of view.

The next application example is to investigate how the chemical reaction rate affects the distribution of product chemical species (i.e. produced new minerals), if all other parameters are kept unchanged in the hydrothermal system considered above. Since the pre-exponential reaction rate constant, to a large extent, represents how fast the chemical reaction proceeds, three different values of the pre-exponential reaction rate constant, namely $10^{-5} \text{kg}/(m^3 \times s)$, $10^{-7} \text{kg}/(m^3 \times s)$ and $10^{-10} \text{kg}/(m^3 \times s)$, have been used in the corresponding computations.

Figures 3.14, 3.15 and 3.16 show the effects of chemical reaction rates on the normalized concentration distributions of reactant and product chemical species in the hydrothermal system. It is obvious that the chemical reaction rate has little influence on either the distribution pattern or the magnitude of the reactant chemical species concentration. Even though the chemical reaction rate may affect the magnitude of the product chemical species concentration, it does not affect the overall distribution pattern of the product chemical species concentration. This finding implies that if the reactant minerals constitute only a small fraction of the whole matrix in a porous medium, which is a commonly accepted assumption in geochemistry (Phillips 1991), the distribution pattern of the normalized concentration of the new mineral produced by chemical reactions is strongly dependent on the characteristics of the convective pore-fluid flow, even though the magnitude of the normalized concentration of the product mineral may also strongly depend on the rates of chemical reactions.

As mentioned previously, any numerical solutions have to be justified before they can be safely accepted and used. For this purpose, we must answer the following questions: Are the numerical results reported in Figs. 3.14, 3.15 and 3.16 correct? Do they represent the physical and chemical characteristics of the hydrothermal system rather than something else (i.e. numerical rubbish)? From these

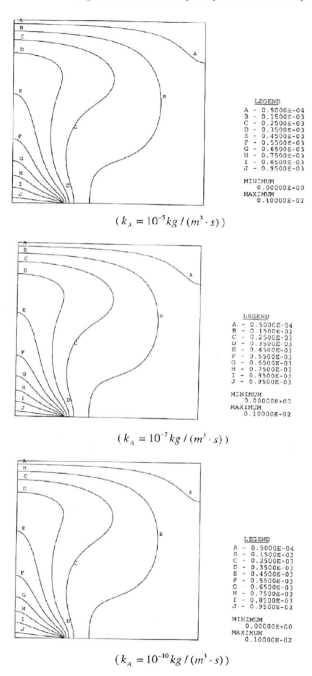

Fig. 3.14 Effects of chemical reaction rates on the normalized concentration distribution of the reactant chemical species (species 1)

3.4 Examples of the Proposed Consistent Point-Searching Interpolation Algorithm   67

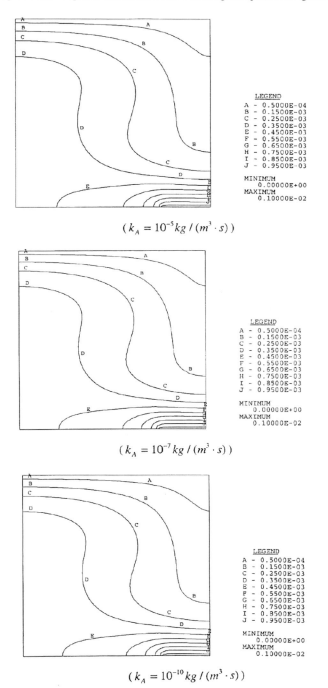

Fig. 3.15 Effects of chemical reaction rates on the normalized concentration distribution of the reactant chemical species (species 2)

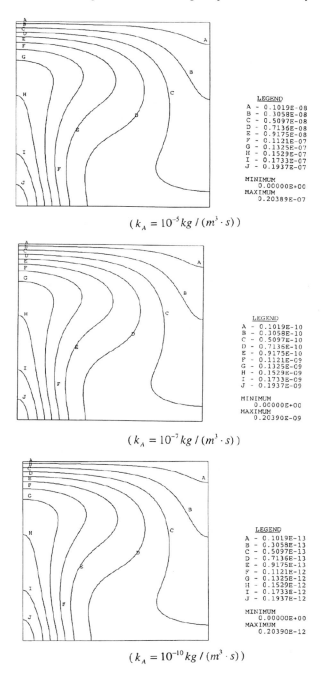

Fig. 3.16 Effects of chemical reaction rates on the normalized concentration distribution of the product chemical species (species 3)

3.4 Examples of the Proposed Consistent Point-Searching Interpolation Algorithm

numerical results, can we find something of more theoretical importance? In the following we will answer the above questions through some theoretical analysis. For the hydrothermal system considered, the maximum change in the density of pore-fluid due to temperature is about 4%, while the maximum change in the density of pore-fluid due to the concentration of any reactant chemical species is only about 0.1%. This means that the buoyancy produced by temperature is much greater than that produced by the concentration of the reactant chemical species. As a result, the convective pore-fluid flow is predominantly driven by the temperature gradient, rather than the concentration gradients of chemical species, even though the contribution of chemical species to the buoyancy is taken into account. More specifically, for the hydrothermal system considered, the velocity components of pore-fluid in the system are strongly dependent on the temperature gradient, rather than the concentration gradients of chemical species. This means that the concentrations of chemical species have minimal influence on the convective pore-fluid flow. Based on this recognition, we can further explore why both the pattern and the magnitude of the normalized concentration of reactant chemical species are very weakly dependent on the rates of the chemical reaction. For the purpose of facilitating the theoretical analysis, we need to rewrite the transport equation (see Eq. (3.5)) for one of the reactant chemical species (i.e. species 1) as follows:

$$\rho_{f0}\left(u\frac{\partial \overline{C}_1}{\partial x} + v\frac{\partial \overline{C}_1}{\partial y}\right) = \rho_{f0}\left(D_{ex}\frac{\partial^2 \overline{C}_1}{\partial x^2} + D_{ey}\frac{\partial^2 \overline{C}_1}{\partial y^2}\right) - \phi k_A \overline{C}_1 \overline{C}_2 \exp\left(\frac{-E_a}{RT}\right). \tag{3.90}$$

Since the reactant chemical species constitutes only a small fraction of the whole matrix in a porous medium, as is commonly assumed in geochemistry (Phillips 1991), we can view the normalized concentration of the reactant chemical species as the first order small quantity in the above equation, at least from the mathematical point of view. Thus, the reaction term in this equation is at least the second order small quantity. This implies that the distribution of the reactant chemical species is controlled by the pore-fluid velocity and the dispersivity of the porous medium, rather than by the chemical reaction term unless the rate of chemical reaction is very fast. This is the reason why both the pattern and the magnitude of the normalized concentration of the reactant chemical species are very weakly dependent on the rates of the chemical reaction in the related numerical results (Figs. 3.14 and 3.15).

Next, we will examine, analytically, why the distribution pattern of the normalized concentration of the product chemical species is almost independent of the rates of the chemical reaction, but the magnitude of the product chemical species is strongly dependent on the rates of the chemical reaction (Fig. 3.16). In this case, we need to rewrite the transport equation for chemical species 3 as follows:

$$\rho_{f0}\left(u\frac{\partial \overline{C}_3}{\partial x} + v\frac{\partial \overline{C}_3}{\partial y}\right) = \rho_{f0}\left(D_{ex}\frac{\partial^2 \overline{C}_3}{\partial x^2} + D_{ey}\frac{\partial^2 \overline{C}_3}{\partial y^2}\right) + \phi k_A \overline{C}_1 \overline{C}_2 \exp\left(\frac{-E_a}{RT}\right). \tag{3.91}$$

Since the magnitude of the normalized concentration of the reactant chemical species is very weakly dependent on the rates of the chemical reaction, the product of $\overline{C}_1$ and $\overline{C}_2$ is almost independent of the chemical reaction rate in Eq. (3.91). As mentioned before, the product of $\overline{C}_1$ and $\overline{C}_2$ is at least the second order small quantity in the theoretical analysis. As a result, the normalized concentration of chemical species 3 is at least the second (or higher) order small quantity. This means that the magnitude of the normalized concentration of chemical species 3 is determined by the reaction term in Eq. (3.91). More specifically, the magnitude of the normalized concentration of chemical species 3 is strongly dependent on the rate of the chemical reaction because the product of ϕ, $\overline{C}_1$ and $\overline{C}_2$ are eventually independent of the chemical reaction rate in Eq. (3.91). If $\overline{C}_{3A}$ is a solution corresponding to the reaction rate constant k_{AA} for Eq. (3.91), we can, mathematically, deduce the solution for another reaction rate constant k_{AB} below.

Since $\overline{C}_{3A}$ is a solution for Equation (3.91), we have the following equation:

$$\rho_{f0}\left(u\frac{\partial \overline{C}_{3A}}{\partial x} + v\frac{\partial \overline{C}_{3A}}{\partial y}\right) = \rho_{f0}\left(D_{ex}\frac{\partial^2 \overline{C}_{3A}}{\partial x^2} + D_{ey}\frac{\partial^2 \overline{C}_{3A}}{\partial y^2}\right) + \phi k_{AA}\overline{C}_1\overline{C}_2 \exp\left(\frac{-E_a}{RT}\right). \tag{3.92}$$

Multiplying Eq. (3.92) by k_{AB}/k_{AA} yields the following equation:

$$\rho_{f0}\frac{k_{AB}}{k_{AA}}\left(u\frac{\partial \overline{C}_{3A}}{\partial x} + v\frac{\partial \overline{C}_{3A}}{\partial y}\right) = \rho_{f0}\frac{k_{AB}}{k_{AA}}\left(D_{ex}\frac{\partial^2 \overline{C}_{3A}}{\partial x^2} + D_{ey}\frac{\partial^2 \overline{C}_{3A}}{\partial y^2}\right) + \phi k_{AB}\overline{C}_1\overline{C}_2 \exp\left(\frac{-E_a}{RT}\right). \tag{3.93}$$

Therefore, the solution corresponding to the reaction rate constant k_{AB} can be straightforwardly expressed as

$$\overline{C}_{3B} = \frac{k_{AB}}{k_{AA}}\overline{C}_{3A}, \tag{3.94}$$

where $\overline{C}_{3B}$ is the solution corresponding to the reaction rate constant k_{AB} for Eq. (3.91).

Equation (3.94) states that for any particular point in the hydrothermal system considered, the normalized concentration of the product chemical species varies linearly with the reaction rate constant. This is the reason why the distribution pattern of concentration of the product chemical species is almost independent of the rates of the chemical reaction, but the magnitude of the product chemical species is strongly dependent on the rates of the chemical reaction (Fig. 3.16). If we further observe the numerical solutions in Fig. 3.16, then we find that the normalized concentration of the product chemical species, indeed, varies linearly with the reaction rate constant. Therefore, the numerical analysis carried out for this application example is further validated by the related theoretical analysis.

It needs to be pointed out that the proposed consistent point-searching interpolation and the related solution methodology have been successfully applied to more

3.4 Examples of the Proposed Consistent Point-Searching Interpolation Algorithm

realistic geological situations (Zhao et al. 1998b, 1999c, Hobbs et al 2000, Gow et al. 2002, Ord et al. 2002, Schaubs and Zhao 2002, Sorjonen-Ward et al. 2002, Zhang et al. 2003), where both complicated geometry and material nonhomogeneity are considered. Although the consistent point-searching interpolation algorithm is developed for the unstructured meshes consisting of 4-node bilinear quadrilateral elements, it can equally be extended to the unstructured meshes consisting of higher order elements. In the latter case, one has to solve the inverse mapping problem numerically, instead of analytically.

Chapter 4
A Term Splitting Algorithm for Simulating Fluid-Rock Interaction Problems in Fluid-Saturated Hydrothermal Systems of Subcritical Zhao Numbers

In recent years, we have been making efforts to develop a practical and predictive tool to explore for giant ore deposits in the upper crust of the Earth. Towards this goal, significant progress has been made towards a better understanding of the basic physical and chemical processes behind ore body formation and mineralization in hydrothermal systems. On the scientific development side, we have developed analytical solutions to answer the following scientific questions (Zhao et al. 1998e, 1999b): (1) Can the pore-fluid pressure gradient be maintained at the value of the lithostatic pressure gradient in the upper crust of the Earth? and, (2) Can convective pore-fluid flow take place in the upper crust of the Earth if there is a fluid/mass leakage from the mantle to the upper crust of the Earth? On the modelling development side, we have developed numerical methods to model the following problems: (1) convective pore-fluid flow in hydrothermal systems (Zhao et al. 1997a, 1998b); (2) coupled reactive pore-fluid flow and species transport in porous media (Zhao et al. 1999a); (3) precipitation and dissolution of minerals in the upper crust of the Earth (Zhao et al. 1998a, 2000a); (4) double diffusion driven pore-fluid flow in hydrothermal systems (Zhao et al. 2000b); (5) pore-fluid flow patterns near geological lenses in hydrodynamic and hydrothermal systems (Zhao et al. 1999d); (6) various aspects of the fully coupled problem involving material deformation, pore-fluid flow, heat transfer and species transport/chemical reactions in pore-fluid saturated porous rock masses (Zhao et al. 1999c, 1999f, 1999g). The above-mentioned work has significantly enriched our knowledge about the physical and chemical processes related to ore body formation and mineralization in the upper crust of the Earth. Since fluid-rock interaction is another potential mechanism of ore body formation and mineralization in the upper crust of the Earth, it is necessary to extend further the developed numerical tools to solve fluid-rock interaction problems. This requires us to deal with coupled reactive species transport phenomenon, which is the direct consequence of the chemical reactions that take place between aqueous reactive species in pore-fluid and solid minerals in pore-fluid saturated porous rock masses.

For a natural fluid-rock interaction system, the reactant chemical species constitutes only a small fraction of the whole matrix in a porous rock (Phillips 1991). In this case, the chemical dissolution front is stable if the Zhao number of the system is subcritical, while it becomes unstable otherwise. Based on the concept of the generalized dimensionless pore-fluid pressure gradient (Zhao et al. 2008e), the corresponding dimensionless Zhao number of a single mineral dissolution system can be defined as follows:

$$Zh = \frac{v_{flow}}{\sqrt{\phi_f D(\phi_f)}} \sqrt{\frac{\overline{V}_p}{k_{chemical} \overline{A}_p C_{eq}}}, \qquad (4.1)$$

where v_{flow} is the Darcy velocity of the pore-fluid flow; (ϕ_f) and $D(\phi_f)$ are the final porosity of the porous medium and the corresponding diffusivity of chemical species after the completion of soluble mineral dissolution; C_{eq} is the equilibrium concentration of the chemical species; $\overline{V}_p$ is the average volume of the soluble grain; $\overline{A}_p$ is the averaged surface area of the soluble grain; $k_{chemical}$ is the rate constant of the chemical reaction.

To understand the physical meanings of each term in the Zhao number, Eq. (4.1) can be rewritten in the following form:

$$Zh = F_{Advection} F_{Diffusion} F_{Chemical} F_{Shape}, \qquad (4.2)$$

where $F_{Advection}$ is a term to represent the solute advection; $F_{Diffusion}$ is a term to represent the solute diffusion/dispersion; $F_{Chemical}$ is a term to represent the chemical kinetics of the dissolution reaction; F_{Shape} is a term to represent the shape factor of the soluble mineral in the fluid-rock interaction system. These terms can be expressed as follows:

$$F_{Advection} = v_{flow}, \qquad (4.3)$$

$$F_{Diffusion} = \frac{1}{\sqrt{\phi_f D(\phi_f)}}, \qquad (4.4)$$

$$F_{Chemical} = \sqrt{\frac{1}{k_{chemical} C_{eq}}}, \qquad (4.5)$$

$$F_{Shape} = \sqrt{\frac{\overline{V}_p}{\overline{A}_p}}. \qquad (4.6)$$

Equations (4.2), (4.3), (4.4), (4.5) and (4.6) clearly indicate that the Zhao number is a dimensionless number that can be used to represent the geometrical, hydrodynamic, thermodynamic and chemical kinetic characteristics of a fluid-rock system in a comprehensive manner. The condition under which a chemical dissolution front in a two-dimensional fluid-saturated porous rock becomes unstable can be expressed by the critical value of this dimensionless number as follows:

$$Zh_{critical} = \frac{(3-\beta)(1+\beta)}{2(1-\beta)}, \qquad \beta = \frac{k(\phi_0)}{k(\phi_f)}, \qquad (4.7)$$

where $Zh_{critical}$ is the critical Zhao number of the fluid-rock interaction system; $k(\phi_0)$ is the initial permeability corresponding to the initial porosity of the porous rock; $k(\phi_f)$ is the final permeability corresponding to the final porosity, ϕ_f, of the porous rock.

Using the concepts of both the Zhao number and the corresponding critical one, the instability of a chemical dissolution front in a fluid-rock interaction system can be determined. The focus of this chapter is to deal with fluid-rock interaction systems of subcritical Zhao numbers, while the focus of the next chapter is to deal with fluid-rock interaction systems of critical and supercritical Zhao numbers.

In terms of numerical modelling of coupled reactive species transport phenomena in pore-fluid saturated porous rock masses, we have divided the reactive transport problems into the following three categories (Zhao et al. 1998a). In the first category of reactive species transport problem, the time scale of the convective/advective flow is much smaller than that of the relevant chemical reaction in porous rock masses so that the rate of the chemical reaction can be essentially taken to be zero in the numerical analysis. For this reason, the first category of species transport problem is often called the non-reactive mass transport problem. In contrast, for the second category of reactive species transport problem, the time scale of the convective/advective flow is much larger than that of the relevant chemical reaction in pore-fluid saturated porous rock masses so that the rate of the chemical reaction can be essentially taken to be infinite, at least from the mathematical point of view. This means that the equilibrium state of the chemical reaction involved is always attained in this category of reactive species transport problem. As a result, the second category of reactive species transport problem is called the quasi-instantaneous equilibrium reaction transport problem. The intermediate case between the first and the second category of reactive species transport problem belongs to the third category of reactive species transport problem, in which the rate of the relevant chemical reaction is a positive real number of finite value. Another significant characteristic of the third category of reactive species transport problem is that the detailed chemical kinetics of chemical reactions must be taken into account. It is the chemical kinetics of a chemical reaction that describes the reaction term in a reactive species transport equation. Due to different regimes in which a chemical reaction proceeds, there are two fundamental reactions, namely homogeneous and heterogeneous reactions, in geochemistry. For homogeneous reactions, the chemical reaction takes place solely between reactive aqueous species. However, for heterogeneous reactions, the chemical reaction takes place at the surfaces between reactive aqueous species and solid minerals. This implies that both solid and fluid phases need to be considered in the numerical modelling of reactive species transport problems with heterogeneous reactions. Although significant achievements have been made for the numerical modelling of non-reactive species and quasi-instantaneous equilibrium reaction transport problems, research on the numerical modelling of the third category of reactive species transport problem is rather limited (Steefel and

Lasaga 1994, Raffensperger and Garven 1995). Considering this fact, we have successfully used the finite element method to model coupled reactive multi-species transport problems with homogeneous reactions (Zhao et al. 1999a). Here we will extend the numerical method developed to model coupled reactive multi-species transport problems with heterogeneous reactions.

4.1 Key Issues Associated with the Numerical Modelling of Fluid-Rock Interaction Problems

Of central importance to the numerical modelling of fluid-rock interaction problems in pore-fluid saturated hydrothermal/sedimentary basins is the appropriate consideration of the heterogeneous chemical reaction which takes place slowly between aqueous reactive species in the pore-fluid and solid minerals in fluid-saturated porous rock masses. From the rock alteration point of view, the pore-fluid flow, which carries reactive aqueous species, is the main driving force causing the heterogeneous chemical reactions at the interface between the pore-fluid and solid minerals so that the rock can be changed from one type into another. The reason for this is that these reactions between the aqueous species and solid minerals may result in dissolution of one kind of mineral and precipitation of another and therefore, the reactive minerals may be changed from one type into another type. Due to the dissolution and precipitation of minerals, the porosity of the porous rock mass evolves during the rock alteration. Since a change in porosity can result in a change in the pore-fluid flow path, a feed-back loop is formed between the pore-fluid flow and the transport of reactive chemical species involved in heterogeneous chemical reactions in fluid-rock interaction systems and this porosity change in the pore-fluid flow needs to be considered in the numerical modelling of fluid-rock interaction problems. This implies that the average linear velocity of pore-fluid flow varies with time due to this porosity evolution. Since both the mesh Peclet number and Courant number are dependent on the average linear velocity, we have to overcome a difficulty in dealing with the problem of variable Peclet and Courant numbers in the transient analysis of fluid-rock interaction problems. Generally, there are two ways to overcome this difficulty, from the computational point of view. The first one is to use a very fine mesh and very small time step of integration so that the requirements for both the mesh Peclet number and Courant number can be satisfied at every time step of the computation. The second one is to regenerate the mesh and re-determine the time step of integration at every time step of computation so that the finite element method can be used effectively. Since either the use of a very fine mesh in the whole process of computation or the regeneration of the mesh at every time step of computation is computationally inefficient, there is a definite need for developing new numerical algorithms to deal with this kind of problem.

Another important issue related to the numerical modelling of fluid-rock interaction problems is that the dissolution rates of minerals are dependent on the existence of the dissolving minerals. Once a dissolving mineral is exhausted in the rock mass, the dissolution rate of this particular mineral must be identically equal to zero. This indicates that the variation in the amount of the dissolving mineral should be considered in the dissolution rates of the minerals. Otherwise, the numerical modelling may violate the real mechanism of the related chemical kinetics and produce incorrect numerical results.

Thus, in order to effectively and efficiently use the finite element method for solving fluid-rock interaction problems in pore-fluid saturated hydrothermal/sedimentary basins, new concepts and numerical algorithms need to be developed for dealing with the following fundamental issues: (1) Since the fluid-rock interaction problem involves heterogeneous chemical reactions between reactive aqueous chemical species in the pore-fluid and solid minerals in the rock masses, it is necessary to develop a new concept involving the generalized concentration of a mineral, so that two types of reactive mass transport equations, namely the conventional mass transport equation for the aqueous chemical species in the pore-fluid and the degenerated mass transport equation for the solid minerals in the rock mass, can be solved simultaneously in computation. (2) Since the reaction area between the pore-fluid and mineral surfaces is basically a function of the generalized concentration of the solid mineral, there is a need to consider the appropriate dependence of the dissolution rate of a dissolving mineral on its generalized concentration in the numerical analysis. (3) Considering the direct consequence of the porosity evolution with time in the transient analysis of fluid-rock interaction problems, the problem of variable mesh Peclet number and Courant number needs to be converted into a problem involving constant mesh Peclet and Courant numbers, so that the conventional finite element method can be directly used to solve fluid-rock interaction problems.

Taking the above-mentioned factors into account, we focus this study on the numerical modelling of mixed solid and aqueous species transport equations with consideration of reaction terms from heterogeneous, isothermal chemical reactions. This means that we will concentrate on the development of new concepts and algorithms so as to solve the fluid-rock interaction problems effectively and efficiently, using the finite element method. For this purpose, a fluid-rock interaction problem, in which K-feldspar ($KAlSi_3O_8$) and/or muscovite ($KAl_3Si_3O_{10}(OH)_2$) are dissolved and muscovite and/or pyrophyllite ($Al_2Si_4O_{10}(OH)_2$) are precipitated, is considered as a representative example in this chapter.

4.2 Development of the Term Splitting Algorithm

For fluid-rock interaction problems in pore-fluid saturated hydrothermal/sedimentary basins, heterogeneous chemical reactions take place at the interface between the reactive aqueous species in the pore-fluid and solid minerals in the rock mass. This means that we need to deal with two types of transport equations in the numerical modelling of fluid-rock interaction problems. The first is the conventional transport equation with the advection/convection term and the diffusion/dispersion term for reactive aqueous species in the pore-fluid. The second is the degenerated transport equation, in which the advection/convection term and the diffusion/dispersion term are identically equal to zero, for solid minerals in the rock mass. In order to solve these two types of transport equations simultaneously, we need to develop the new concept of the generalized concentration for the solid mineral. The generalized concentration of a solid mineral is defined as the moles of the solid mineral per unit total volume (i.e. the volume of void plus the volume of solid particles) of the porous rock mass. Using this new concept, the general form of the second type of transport equation to be solved in this study can be expressed as follows:

$$\frac{\partial C_{Gi}}{\partial t} = \phi R_i \quad (i = 1, 2, \ldots, m), \quad (4.8)$$

where C_{Gi} is the generalized concentration of solid mineral/species i; ϕ is the porosity of the porous rock mass; R_i is the reaction rate of solid mineral/species i and m is the number of the solid minerals to be considered in the system.

For the reactive aqueous species, the general form of the first type of transport equation in a two dimensional pore-fluid saturated, isotropic and homogeneous porous medium reads

$$\phi \frac{\partial C_{Ci}}{\partial t} + u_x \frac{\partial C_{Ci}}{\partial x} + u_y \frac{\partial C_{Ci}}{\partial y} = D_i^e \nabla^2 C_{Ci} + \phi R_i \quad (i = m+1, m+2, \ldots, n), \quad (4.9)$$

where C_{Ci} is the conventional concentration of aqueous species i; u_x and u_y are the Darcy velocities of pore-fluid in the x and y directions respectively; D_i^e is of the following form:

$$D_i^e = \phi D_i, \quad (4.10)$$

where D_i is the dispersivity of aqueous species i.

It needs to be pointed out that under some circumstances, where the reactant chemical species constitute only a small fraction of the whole matrix in a porous medium (Phillips 1991), the total flux of pore-fluid flow in a horizontal aquifer may be approximately considered as a constant. For example, in a groundwater supply system, groundwater can be pumped out from a horizontal aquifer at a constant flow rate. In a geological system, topographically induced pore-fluid flow can also flow through a horizontal aquifer at a constant flow rate. This indicates that in order to satisfy the mass conservation requirement of the pore-fluid in the above-mentioned aquifers, the total flux of the pore-fluid flow should be constant through all vertical cross-sections, which are perpendicular to the direction of pore-fluid flow. Therefore, the Darcy velocity in the flow direction is constant if the horizontal aquifer has a constant thickness. The reason for this is that the Darcy velocity is the velocity averaged over the total area of a representative elementary area (Zhao et al. 1998e, Phillips 1991, Nield and Bejan 1992, Zhao et al. 1994c), rather than over the pore area of the representative elementary area. In other words, the Darcy velocity in the above-mentioned aquifers can be maintained at a constant value, even though

4.2 Development of the Term Splitting Algorithm

the vertical cross-sections, which are perpendicular to the direction of the pore-fluid flow, may have different values of the areal porosity (Zhao 1998e, Phillips 1991, Nield and Bejan 1992, Zhao et al. 1994c). However, the linear averaged velocity, which is the velocity averaged over the pore area of a representative elementary area, varies if the vertical cross-sections have different values of the areal porosity.

It is also noted that Eq. (4.9) is a typical transport equation for species i. This equation describes the mass conservation of species i in the pore-fluid saturated porous medium for both constant and variable porosity cases (Zhao et al. 1998e, 1994c).

In order to consider the feedback effect of porosity due to heterogeneous chemical reactions, we need a porosity evolution equation in the numerical analysis. Using the generalized concentration of solid minerals, the evolution equation for the porosity of the rock mass can be expressed as follows:

$$\phi = 1 - \sum_{i=1}^{m} \frac{W_i}{\rho_s} C_{Gi}, \qquad (4.11)$$

where W_i is the molecular weight of mineral i; ρ_s is the unit weight of solid minerals. For the transient analysis, C_{Gi} are the functions of space and time variables, so that the porosity of the rock mass varies with both space and time. This is the reason why Eq. (4.11) is called the evolution equation of porosity with time.

From the chemical kinetics point of view, the reaction rates involved in Eqs. (4.8) and (4.9) can be expressed as follows (Lasaga 1984):

$$R_i = -\sum_{j=1}^{N} \tau_{ij} r_j, \qquad (4.12)$$

where τ_{ij} is the contribution factor of mineral/species j to the reaction rate of mineral/species i; N is the number of the heterogeneous chemical reactions to be considered in the rock mass; r_j is the overall dissolution rate of mineral j.

$$r_j = \frac{A_j}{V_f} k_j \left(1 - \frac{Q_j}{K_j}\right), \qquad (4.13)$$

where A_j is the surface area of mineral j; k_j and K_j are the reaction constant and the equilibrium constant of the jth heterogeneous chemical reaction; V_f is the volume of the solution; Q_j is the chemical affinity of the jth heterogeneous chemical reaction.

It needs to be pointed out that the surface area of a dissolving mineral varies in the process of mineral dissolution. Once the dissolving mineral is depleted in the rock mass, the surface area of the mineral must be equal to zero and therefore, the chemical reaction dissolving this particular mineral should stop. If this area is simply assumed to be a constant in the numerical modelling, then the resulting numerical solutions are incorrect because the real mechanism of chemical kinetics cannot be simulated during the consumption of the dissolving mineral in the rock mass.

In other words, the use of a constant surface area cannot simulate the dependent nature of the chemical reaction rate on a change in the amount of the dissolving mineral during the heterogeneous chemical reaction process. In addition, the use of a constant surface area cannot automatically terminate the dissolution reaction, even though the dissolving mineral is absolutely exhausted in the rock mass. Since the surface area of a dissolving mineral is strongly dependent on its generalized concentration, we can establish the following relation between the surface area and the generalized concentration of the dissolving mineral:

$$\frac{A_j}{V_f} = \alpha_j C_{Gj}^q, \quad (4.14)$$

where α_j and q are positive real numbers.

Like the surface area of the dissolving minerals, the values of both α_j and q are dependent on the constituents, packing form, grain size and so forth of the minerals. For simple packing of loose uniform particles of the minerals, these values can be determined analytically. For example, in the case of packing uniform circles in a two dimensional domain, $q = 1/2$, whereas in the case of packing uniform spheres in a three dimensional domain, $q = 2/3$. However, for a real rock mass in pore-fluid saturated hydrothermal/sedimentary basins, the values of both α_j and q need to be determined by field measurements and laboratory tests.

Inserting Eq. (4.14) into Eq. (4.13) yields the following equation:

$$r_j = \alpha_j C_{Gj}^q k_j \left(1 - \frac{Q_j}{K_j}\right). \quad (4.15)$$

Clearly, this equation states that once the dissolving mineral is gradually consumed in the rock mass, the generalized concentration of this mineral continuously evolves to zero and therefore, its dissolution rate is automatically set to be zero when it is completely depleted in the numerical analysis. This is the first advantage of introducing the new concept of the generalized concentration for solid minerals.

Another advantage of introducing the concept of the generalized concentration for solid minerals is that both the first type and second type of transport equations (i.e. Eqs. (4.8) and (4.9)) can be solved simultaneously. Since several heterogeneous chemical reactions take place simultaneously in dissolution problems involving multiple minerals, it is important to solve simultaneously all the transport equations with heterogeneous reaction terms in fluid-rock interaction problems, if we want to simulate the chemical kinetics of these heterogeneous chemical reactions correctly.

If the transient process of a heterogeneous chemical reaction is of interest, then the dissolution and precipitation of minerals can result in the variation of porosity with time. This indicates that the linear average velocity of pore-fluid, which is involved in the advection/convection term of a reactive aqueous species transport equation, varies with time during the transient analysis. Even for a horizontal aquifer, in which the horizontal Darcy velocity may be constant, the related average linear velocity still varies with time, because it is inversely proportional

4.2 Development of the Term Splitting Algorithm

to the porosity of the aquifer. In this regard, we have to deal with the problem of variable Peclet and Courant numbers in the transient analysis of fluid-rock interaction problems. To solve this problem effectively and efficiently, we develop the term splitting algorithm in this section. The basic idea behind this algorithm is that through some mathematical manipulations, we invert the problem of variable Peclet and Courant numbers into that of constant Peclet and Courant numbers so that a fixed finite element mesh and fixed time step of integration can be safely used in the numerical modelling of fluid-rock interaction problems.

If the pore-fluid saturated porous medium has a non-zero initial porosity, then we can use this initial porosity as a reference porosity and change Eq. (4.9) into the following form:

$$\frac{\phi}{\phi_0}\phi_0\frac{\partial C_{Ci}}{\partial t} + u_x\frac{\partial C_{Ci}}{\partial x} + u_y\frac{\partial C_{Ci}}{\partial y} = D_i^e \nabla^2 C_{Ci} + \frac{\phi}{\phi_0}\phi_0 R_i \quad (i = m+1, m+2, \ldots, n). \tag{4.16}$$

Dividing both the left and right hand sides of Eq. (4.16) by ϕ/ϕ_0 yields the following equation:

$$\phi_0\frac{\partial C_{Ci}}{\partial t} + \frac{\phi_0}{\phi}u_x\frac{\partial C_{Ci}}{\partial x} + \frac{\phi_0}{\phi}u_y\frac{\partial C_{Ci}}{\partial y} = \frac{\phi_0}{\phi}D_i^e \nabla^2 C_{Ci} + \phi_0 R_i \quad (i = m+1, m+2, \ldots, n). \tag{4.17}$$

Using the initial Darcy velocities, u_{x0} and u_{y0}, as reference velocities in the x and y directions respectively, Eq. (4.17) can be rewritten as follows:

$$\phi_0\frac{\partial C_{Ci}}{\partial t} + \left[u_{x0} + \left(\frac{\phi_0}{\phi}u_x - u_{x0}\right)\right]\frac{\partial C_{Ci}}{\partial x} + \left[u_{y0} + \left(\frac{\phi_0}{\phi}u_y - u_{y0}\right)\right]\frac{\partial C_{Ci}}{\partial y}$$
$$= \frac{\phi_0}{\phi}D_i^e \nabla^2 C_{Ci} + \phi_0 R_i$$
$$(i = m+1, m+2, \ldots, n). \tag{4.18}$$

It is noted that in Eq. (4.18), the coefficient in front of $\partial C_{Ci}/\partial x$ has been split into a constant coefficient, u_{x0}, and a variable term, $(u_x\phi_0)/\phi - u_{x0}$. Similarly, the coefficient in front of $\partial C_{Ci}/\partial y$ has also been split into a constant coefficient, u_{y0}, and a variable term, $(u_y\phi_0)/\phi - u_{y0}$. This term splitting process, although it is carried out mathematically, forms a very important step in the proposed term splitting algorithm.

If the first derivative terms with constant coefficients are kept in the left hand side of Eq. (4.18) and the rest terms are moved to the right hand side, Eq. (4.18) can be rewritten as follows:

$$\phi_0\frac{\partial C_{Ci}}{\partial t} + u_{x0}\frac{\partial C_{Ci}}{\partial x} + u_{y0}\frac{\partial C_{Ci}}{\partial y} = D_{0i}^e \nabla^2 C_{Ci} + \phi_0 R_i + R_i^e \quad (i = m+1, m+2, \ldots, n), \tag{4.19}$$

$$R_i^e = \left(u_{x0} - \frac{\phi_0}{\phi} u_x\right) \frac{\partial C_{Ci}}{\partial x} + \left(u_{yo} - \frac{\phi_0}{\phi} u_y\right) \frac{\partial C_{Ci}}{\partial y} \quad (i = m+1, m+2, \ldots, n), \tag{4.20}$$

$$D_{0i}^e = \phi_0 D_i \quad (i = m+1, m+2, \ldots, n), \tag{4.21}$$

where ϕ_0 is the initial porosity of the porous rock mass; u_{x0} and u_{y0} are the reference velocities, which are the initial Darcy velocities, in the x and y directions; R_i^e is the equivalent source/sink term due to the variation in both the velocity of pore-fluid and the porosity of the rock mass.

It is clear that the proposed term splitting algorithm consists of the following two main steps. First, by means of some rigorous mathematical manipulations, Eq. (4.9) with variable coefficients in front of the first and second derivatives has been changed into Eq. (4.19), which has constant coefficients in front of the first and second derivatives. From the mathematical point of view, this means that we have changed a partial differential equation with variable coefficients (Eq. (4.9)) into another one with constant coefficients in front of the first and second derivatives. Second, Eq. (4.19) is directly used to obtain a numerical solution for Equation (4.9) in the finite element analysis, because both Eqs. (4.19) and (4.9) are mathematically equivalent. However, from the computational point of view, Eq. (4.19) can be solved much easier than Eq. (4.9) in the finite element analysis. The reason for this is that in order to solve Eq. (4.9), we need to deal with a problem with variable mesh Peclet number and Courant number. But in order to solve Eq. (4.19), we need only to deal with a problem with constant mesh Peclet number and Courant number. On the other hand, the numerical solvers presently available are more stable and robust when they are used to solve Eq. (4.19), instead of Eq. (4.9).

Note that the equivalent source/sink term presented here has a very clear physical meaning. For a representative elementary volume/area, a change either in the porosity or in the pore-fluid velocity is equivalent to the addition of a source/sink term into the representative elementary volume/area. Since the constants involved in both the advection and dispersion terms are basically independent of time, it is possible to use the initial mesh and time step, which are determined using the initial conditions at the beginning of computation, throughout the whole process of the transient analysis of fluid-rock interaction problems. As a result, the finite element method with the proposed term splitting algorithm can be efficiently used to solve both the first and second type of transient transport equations (i.e. Eqs. (4.8) and (4.19)) for fluid-rock interaction problems in pore-fluid saturated hydrothermal/sedimentary basins.

4.3 Application Examples of the Term Splitting Algorithm

In order to illustrate the usefulness and applicability of the newly proposed concepts and numerical algorithms, we have built them into a finite element code so that fluid-rock interaction problems in pore-fluid saturated hydrothermal/sedimentary basins can be solved effectively and efficiently. As shown in Fig. 4.1, the application exam-

4.3 Application Examples of the Term Splitting Algorithm

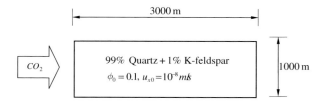

Fig. 4.1 Geometry and initial conditions for the fluid-rock interaction problem in a pore-fluid saturated aquifer

ple considered in this section is an isothermal fluid-rock interaction problem in a pore-fluid saturated horizontal aquifer within a sedimentary basin. The topographically induced pore-fluid flow is horizontally from the left to the right of the aquifer. This means that the horizontal Darcy velocity is constant within the aquifer. The rock of the aquifer is initially composed of K-feldspar ($KAlSi_3O_8$) and quartz (SiO_2). There is an injection of carbon dioxide gas (CO_2) at the left inlet of the aquifer. The injected carbon dioxide flow in the aquifer may dissolve K-feldspar and precipitate muscovite ($KAl_3Si_3O_{10}(OH)_2$). If the injected carbon dioxide flow is strong enough, the precipitated muscovite may be dissolved and pyrophyllite ($Al_2Si_4O_{10}(OH)_2$) will be precipitated. Basically, the following three overall chemical reactions may take place in the aquifer.

$$CO_2 + H_2O \stackrel{fast}{\Longrightarrow} HCO_3^- + H^+, \tag{4.22}$$

$$3KAlSi_3O_8 + 2H^+ \stackrel{k_1}{\Longrightarrow} 2K^+ + KAl_3Si_3O_{10}(OH)_2 + 6SiO_2, \tag{4.23}$$

$$2KAl_3Si_3O_{10}(OH)_2 + 2H^+ + 6SiO_2 \stackrel{k_2}{\Longrightarrow} 2K^+ + 3Al_2Si_4O_{10}(OH)_2. \tag{4.24}$$

The first reaction (i.e. Eq. (4.22)) states that when the injected carbon dioxide (CO_2) gas enters the fluid-rock system, it reacts very fast with water (H_2O), so that the chemical equilibrium can be reached quasi-instantaneously. This means that the injection of carbon dioxide (CO_2) gas is equivalent to the direct injection of H^+ into the system. As will be demonstrated later, the use of this equivalence may result in a reduction in the number of primary reactive chemical species and therefore, a considerable reduction in the degrees of freedom of the whole system. As a direct consequence, the number of reactive species transport equations can be reduced to the minimum in computation. This can lead to a significant reduction in requirements of both the computer storage and CPU time in the numerical modelling of fluid-rock interaction problems.

It needs to be pointed out that Eq. (4.23) describes a heterogeneous chemical reaction, in which K-feldspar ($KAlSi_3O_8$) is dissolved and muscovite ($KAl_3Si_3O_{10}(OH)_2$) is precipitated, while Eq. (4.24) describes a heterogeneous chemical reaction, in which muscovite ($KAl_3Si_3O_{10}(OH)_2$) is dissolved and pyrophyllite ($Al_2Si_4O_{10}(OH)_2$) is precipitated in the fluid-rock system. Chemically,

Eq. (4.23) states that the dissolution of one mole of K-feldspar needs to consume 2/3 moles of H^+ and then produces 1/3 moles of muscovite, 2/3 moles of K^+ and two moles of quartz. Similarly, Eq. (4.24) states that the dissolution of one mole of muscovite needs to consume one mole of H^+ and two moles of quartz and then produces 1.5 moles of pyrophyllite and one mole of K^+. This indicates that we only need to consider six primary chemical reactive species, namely two aqueous species (K^+ and H^+) in the pore-fluid and four solid minerals/species (K-feldspar, muscovite, pyrophyllite and quartz) in the rock mass, in the computation.

As shown in Fig. 4.1, the computational domain to be used in the numerical analysis is a rectangle of 3000 m by 1000 m in size. This computational domain is discretized into 2700 4-node quadrilateral elements. Since the problem to be considered here is essentially an initial value problem, the following initial conditions are used in the computation. The initial porosity of the porous medium is 0.1. The initial values of the generalized concentrations of quartz and K-feldspar are 36.82 kmol/m^3 and 0.87 kmol/m^3 respectively. It is assumed that all the aqueous reactive species involved in the chemical reactions are in chemical equilibrium at the beginning of the computation. Under this assumption, the initial values of the conventional concentrations of K^+ and H^+ are 0.1 kmol/m^3 and 1.6×10^{-5} kmol/m^3. The horizontal Darcy velocity of pore-fluid flow is 10^{-8} m/s in the horizontal aquifer. The dispersion coefficient of K^+ is 2×10^{-6} m^2/s. Since the injected carbon dioxide (CO_2) gas diffuses much faster than aqueous species, the dispersion coefficient of the injected H^+, which is the equivalence of the injected carbon dioxide (CO_2) gas, is 100 times that of K^+ in the computation. The concentration of the injected H^+ is 6.4×10^{-3} kmol/m^3 at the left vertical boundary of the computational domain. For the purpose of simulating the chemical kinetics of heterogeneous reactions, the following thermodynamic data are used. The chemical equilibrium constants are 3.89×10^7 and 8318.0 for the K-feldspar and muscovite dissolution reactions, whereas the chemical reaction rate constants for these two dissolution reactions are 5.03×10^{-12} kmol/(m^{2}s) and 4.48×10^{-12} kmol/(m^{2}s) respectively. α_j ($j = 1, 2$) and q are assumed to be unity and 0.5 in the numerical analysis. In addition, the integration time step is set to be 3×10^8 s, which is approximately equal to 10 years, in the computation.

Figures 4.2, 4.3 and 4.4 show the generalized concentration distributions of K-feldspar, muscovite and pyrophyllite in the fluid-rock interaction system at four different time instants, namely $t = 3 \times 10^9$s, $t = 1.5 \times 10^{10}$s, $t = 6 \times 10^{10}$s and $t = 1.5 \times 10^{11}$s respectively. It is observed the dissolution front (i.e. the region from red to blue in Fig. 4.2) of K-feldspar propagates from the left side to the right side of the computational domain. Since muscovite can be precipitated and dissolved at the same time, there are two propagation fronts, the precipitation front and dissolution front, in Fig. 4.3. The precipitation front of muscovite is the region from blue to red in the right half of the computational domain, whereas the dissolution front is the region from red to blue in the left half of the computational domain. Both the precipitation and dissolution fronts propagate from the left side to the right side of the computational domain. In the case of the generalized concentration distribution of pyrophyllite (see Fig. 4.4), there only exists the precipitation front (i.e. the region

4.3 Application Examples of the Term Splitting Algorithm

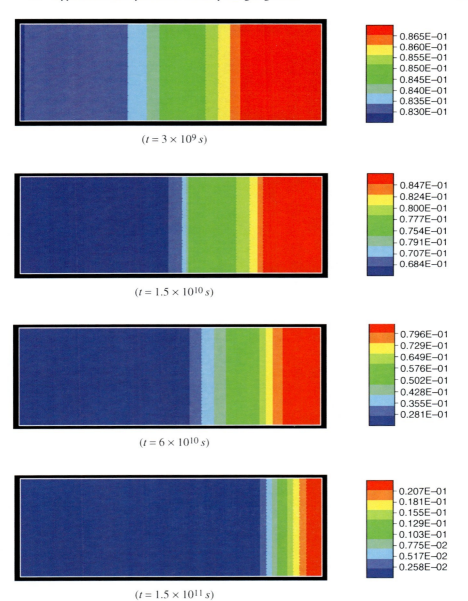

Fig. 4.2 Distribution of K-feldspar in the fluid-rock interaction system

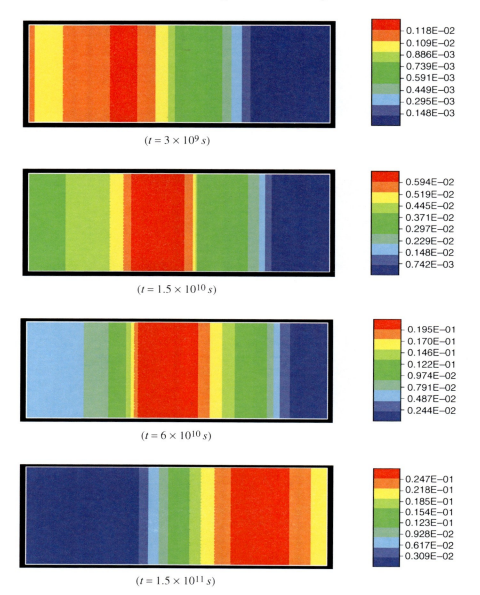

Fig. 4.3 Distribution of muscovite in the fluid-rock interaction system

4.3 Application Examples of the Term Splitting Algorithm

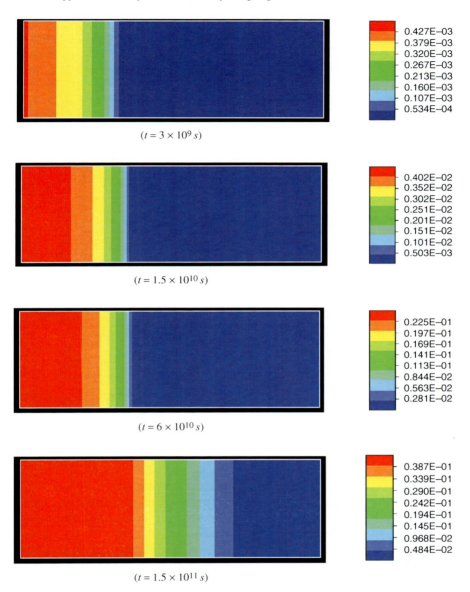

Fig. 4.4 Distribution of pyrophyllite in the fluid-rock interaction system

from blue to red in Fig. 4.4) of pyrophyllite in the computational domain. Clearly, all the dissolution/precipitation propagation fronts propagate from the left side to the right side of the computational domain, which is exactly in the same direction as the pore-fluid flow in the aquifer. This indicates that when the injected carbon dioxide (CO_2) gas enters the fluid-rock interaction system, it produces H^+ quasi-instantaneously at the left entrance of the system. The produced H^+ propagates from the left side to the right side of the aquifer so that K-feldspar gradually dissolves and muscovite precipitates along the same direction as the dissolution front propagation of K-feldspar. Since the dissolution reaction constants of K-feldspar and muscovite are of the same order in magnitude, the dissolution process of muscovite is in parallel with that of K-feldspar. This is to say, the precipitated muscovite from the dissolution of K-feldspar can be dissolved and therefore, pyrophyllite can be precipitated at an early stage, as clearly exhibited in Fig. 4.4. This fact indicates that in order to model the chemical kinetics of the involved heterogeneous reactions correctly, all the reactive species transport equations should be solved simultaneously in the numerical analysis.

It is noted that from the mathematical point of view, the problem solved here is an initial value problem, rather than a boundary value problem. For an initial value problem in a homogeneous, isotropic porous medium (as we considered here), the chemical reaction/propagation front of chemically reactive species is stable before the system reaches a steady state, from both the physical and chemical points of view. However, if the numerical algorithm is not robust enough and the mesh/time step used is not appropriate in a numerical analysis, numerical error may result in an unstable/oscillatory chemical reaction/propagation front (Zienkiewicz 1977). Just as Phillips (1991) stated, "Numerical calculations can converge to a grid-dependent limit, and artifacts of a solution can be numerical rather than geological". This indicates that any numerical solution must be validated, at least qualitatively if the analytical solution to the problem is not available. We emphasise the importance of this issue here because it is often overlooked by some purely numerical modellers. The simplest way to evaluate a numerical solution in a qualitative manner is to check whether or not the solution violates the common knowledge related to the problem studied. Since all the propagation fronts of chemically reactive species in Figs. 4.2, 4.3 and 4.4 are comprised of vertically parallel lines, they agree very well with common knowledge, as the chemical dissolution system considered here is subcritical. This demonstrates that the proposed numerical algorithms are robust enough for solving fluid-rock interaction problems, at least from a qualitative point of view.

Figures 4.5, 4.6 and 4.7 show the conventional/generalized concentration distributions of K^+, H^+ and quartz in the fluid-rock interaction system at four different time instants, namely $t = 3 \times 10^9$s, $t = 1.5 \times 10^{10}$s, $t = 6 \times 10^{10}$s and $t = 1.5 \times 10^{11}$s respectively. Generally, K^+ is gradually produced and accumulated with time due to both the K-feldspar and muscovite dissolution reactions, whereas H^+ is continuously injected at the left entrance of the aquifer and consumed with time due to both the K-feldspar and muscovite dissolution reactions. Quartz is produced by the K-feldspar dissolution reaction but is consumed by the muscovite dissolution reaction. Since the generalized concentration of K-feldspar is greater than that of

4.3 Application Examples of the Term Splitting Algorithm

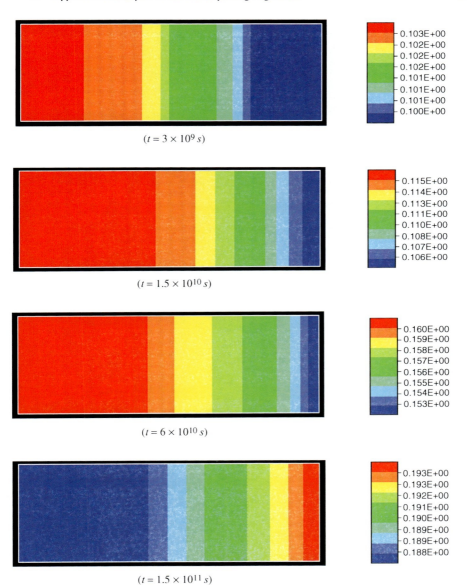

Fig. 4.5 Distribution of K^+ in the fluid-rock interaction system

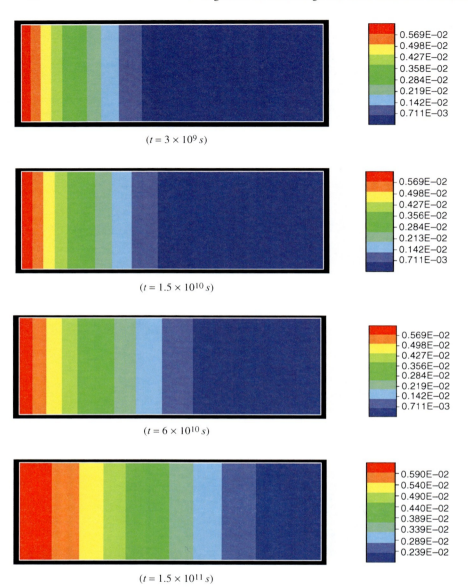

Fig. 4.6 Distribution of H^+ in the fluid-rock interaction system

4.3 Application Examples of the Term Splitting Algorithm

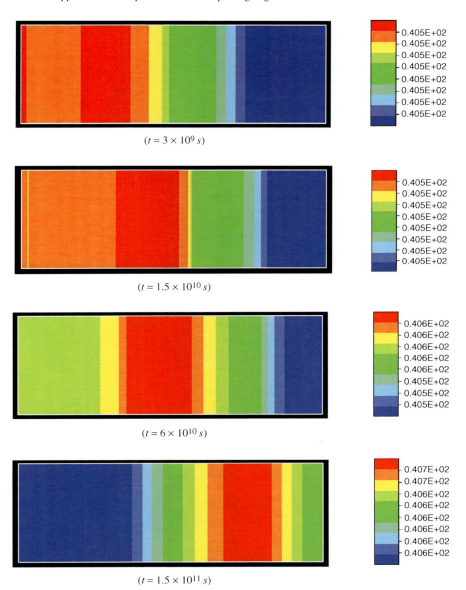

Fig. 4.7 Distribution of quartz in the fluid-rock interaction system

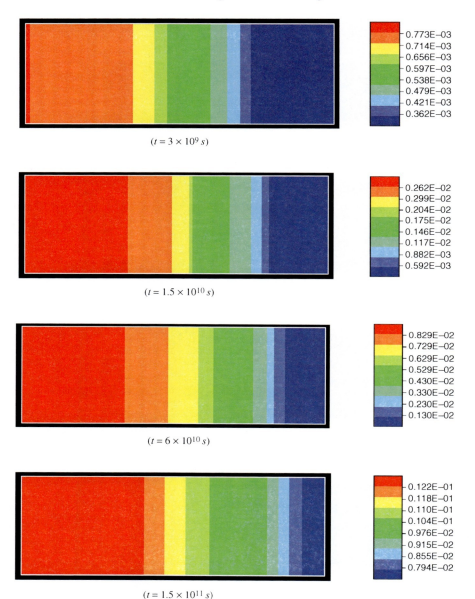

Fig. 4.8 Distribution of porosity variation in the fluid-rock interaction system

muscovite within the time frame considered, the accumulation of quartz exceeds the consumption of quartz and therefore, the generalized concentration of quartz increases with time due to both the K-feldspar and muscovite dissolution reactions. In the case of the conventional concentration of K^+, its maximum values are 0.1032 kmol/m^3 and 0.1942 kmol/m^3 at $t = 3 \times 10^9$s and $t = 1.5 \times 10^{11}$s respectively. This indicates a significant increase in the conventional concentration of K^+ with the increase of time within the scope of this study. However, in the case of the conventional concentration of H^+, its maximum value is a constant of 6.4×10^{-3}kmol/m^3 at both $t = 3 \times 10^9$s and $t = 1.5 \times 10^{11}$s. The reason for this is due to a continuous injection of H^+ at the left entrance of the aquifer. In the case of the generalized concentration of quartz, its maximum values are 0.4051 kmol/m^3 and 0.4067 kmol/m^3 at $t = 3 \times 10^9$s and $t = 1.5 \times 10^{11}$s respectively. This indicates a slight increase in the generalized concentration of quartz with the increase of time within the scope of this study.

Figure 4.8 shows the porosity variation, $(\phi - \phi_0)/\phi_0$, in the fluid-rock interaction system at four different time instants. It is clear that the porosity of the porous medium evolves with time in the process of fluid-rock interactions. The evolution of porosity mainly depends on the evolution of the K-feldspar dissolution, muscovite precipitation and dissolution and pyrophyllite precipitation in the fluid-rock interaction system. In addition, the porosity variation front propagates from the left side to the right side of the aquifer, which is identical to the direction of pore-fluid flow in the aquifer. The propagation of the porosity variation front can be clearly observed from the numerical results shown in Fig. 4.8.

In summary, the related numerical solutions from an application example, which is a K-feldspar dissolution problem in a pore-fluid saturated, isothermal and homogeneous aquifer, have demonstrated that: (1) There exist only a dissolution propagation front for K-feldspar and a precipitation propagation front for pyrophyllite, but there exist a precipitation propagation front and a dissolution propagation front for muscovite during the heterogeneous chemical reactions in the aquifer. (2) The dissolution of K-feldspar and muscovite may take place simultaneously in the aquifer so that pyrophyllite can be precipitated at the early stage of the heterogeneous chemical reactions. (3) All the propagation fronts of chemically reactive species are comprised of vertically parallel lines, the propagation directions of which are exactly the same as that of the pore-fluid flow in the aquifer. (4) The evolution of porosity mainly depends on the evolution of K-feldspar dissolution, muscovite precipitation and dissolution and pyrophyllite precipitation in the fluid-rock interaction system.

Chapter 5
A Segregated Algorithm for Simulating Chemical Dissolution Front Instabilities in Fluid-Saturated Porous Rocks

When fresh pore-fluid flow enters a solute-saturated porous medium, where the concentration of the solute (i.e. aqueous mineral) reaches its equilibrium concentration, the concentration of the aqueous mineral is diluted so that the solid part of the solute (i.e. solid mineral) is dissolved to maintain the equilibrium state of the solution. This chemical dissolution process can result in the propagation of a dissolution front within the fluid-saturated porous medium. Due to the dissolution of the solid mineral, the porosity of the porous medium is increased behind the dissolution front. Since a change in porosity can cause a remarkable change in permeability, there is a feedback effect of the porosity change on the pore-fluid flow, according to Darcy's law. It is well known that because pore-fluid flow plays an important role in the process of reactive chemical-species transport, a change in pore-fluid flow can cause a considerable change in the chemical-species concentration within the porous medium (Steefel and Lasaga 1990, 1994, Yeh and Tripathi 1991, Raffensperger and Garven 1995, Shafter et al. 1998a, b, Xu et al. 1999, 2004, Ormond and Ortoleva 2000, Chen and Liu 2002, Zhao et al. 2005a, 2006c). This means that the problem associated with the propagation of a dissolution front is a fully coupled nonlinear problem between porosity, pore-fluid pressure and reactive chemical-species transport within the fluid-saturated porous medium. If the fresh pore-fluid flow is slow, the feedback effect of the porosity change is weak so that the dissolution front is stable. However, if the fresh pore-fluid flow is fast enough, the feedback effect of the porosity change becomes strong so that the dissolution front becomes unstable. In this case, a new morphology (i.e. dissipative structure) of the dissolution front can emerge due to the self-organization of this coupled nonlinear system. This leads to an important scientific problem, known as the reactive infiltration instability problem (Chadam et al. 1986, 1988, Ortoleva et al. 1987), which is closely associated with mineral dissolution in a fluid-saturated porous medium.

This kind of chemical-dissolution-front instability problem exists ubiquitously in many scientific and engineering fields. For example, in geo-environmental engineering, the rehabilitation of contaminated sites using fresh water to wash the sites involves the propagation problem of the removed contaminant material front in water-saturated porous medium. In mineral mining engineering, the extraction of minerals in the deep Earth using the in-situ leaching technique may result in the propagation problem of the dissolved mineral front in fluid-saturated porous medium. In the petroleum industry, the secondary recovery of oil by acidifying the oil field to uniformly increase porosity and hence the yield of oil is associated with the propagation of the acid-dissolved material front in porous rocks. More importantly, due to the ever-increasing demand for mineral resources and the likelihood of the exhaust of the existing ore deposits, it is imperative to develop advanced techniques to explore for new ore deposits. Towards this goal, there is a definite need to understand the important physical and chemical processes that control ore body formation and mineralization in the deep Earth (Raffensperger and Garven 1995, Zhao et al. 1997a, 1998a, 1999b, d, 2000b, 2001b, d, Gow et al. 2002, Ord et al. 2002, Schaubs and Zhao 2002, Zhao et al. 2002c, 2003a). According to modern mineralization theory, ore body formation and mineralization is mainly controlled by pore-fluid flow focusing and the equilibrium concentration gradient of the concerned minerals (Phillips 1991, Zhao et al. 1998a). Since the chemical dissolution front can create porosity and therefore can locally enhance the pore-fluid flow, it becomes a potentially powerful mechanism to control ore body formation and mineralization in the deep Earth.

Although analytical solutions can be obtained for some reactive transport problems with simple geometry, it is very difficult, if not impossible, to predict analytically the complicated morphological evolution of a chemical dissolution front in the case of the chemical dissolution system becoming supercritical. As an alternative, numerical methods are suitable to overcome this difficulty. Since numerical methods are approximate solution methods, they must be verified before they are used to solve any new type of scientific and engineering problem. For this reason, it is necessary to derive the analytical solution for the propagation of a planar dissolution front within a benchmark problem, the geometry of which can be accurately simulated using numerical methods such as the finite element method (Zienkiewicz 1977, Lewis and Schrefler 1998) and the finite difference method. This makes it possible to compare the numerical solution obtained from the benchmark problem with the derived analytical solution so that the proposed numerical procedure can be verified for simulating chemical-dissolution-front propagation problem in the fluid-saturated porous medium.

5.1 Mathematical Background of Chemical Dissolution Front Instability Problems in Fluid-Saturated Porous Rocks

5.1.1 A General Case of Reactive Multi-Chemical-Species Transport with Consideration of Porosity/Permeability Feedback

For a pore-fluid-saturated porous medium, Darcy's law can be used to describe pore-fluid flow and Fick's law can be used to describe mass transport phenomena respectively. If both the porosity change of the porous medium is caused by

5.1 Mathematical Background of Chemical Dissolution Front Instability Problems

chemical dissolution of soluble solid minerals within the porous medium and the feedback effect of such a change on the variation of permeability and diffusivity are taken into account, the governing equations of the coupled nonlinear problem between porosity, pore-fluid flow and reactive multi-chemical-species transport in the pore-fluid-saturated porous medium can be expressed as follows:

$$\frac{\partial}{\partial t}(\rho_f \phi) + \nabla \bullet (\rho_f \phi \vec{u}_{linear}) = 0, \tag{5.1}$$

$$\vec{u} = \phi \vec{u}_{linear} = -\frac{k(\phi)}{\mu} \nabla p, \tag{5.2}$$

$$\frac{\partial}{\partial t}(\phi C_i) + \nabla \bullet (\phi C_i \vec{u}_{linear}) = \nabla \bullet [\phi D_i(\phi) \nabla C_i] + R_i \qquad (i = 1, 2, \ldots N), \tag{5.3}$$

where $\vec{u}_{linear}$ is the averaged linear velocity vector within the pore space of the porous medium; $\vec{u}$ is the Darcy velocity vector within the porous medium; p and C_i are pressure and the concentration (moles/pore-fluid volume) of chemical species i; μ is the dynamic viscosity of the pore-fluid; ϕ is the porosity of the porous medium; $D_i(\phi)$ is the diffusivity of chemical species i; ρ_f is the density of the pore-fluid; N is the total number of all the chemical species to be considered in the system; R_i is the source/sink term of chemical species i due to the dissolution/precipitation of solid minerals within the system; $k(\phi)$ is the permeability of the porous medium.

It is noted that in Eqs. (5.1), (5.2) and (5.3), the chemical species concentration, the fluid density and averaged linear velocity of the pore-fluid are defined in the pore space, while the source/sink term and the Darcy velocity of the pore fluid are defined in the whole medium space (Phillips 1991, Nield and Bejan 1992, Zhao et al. 1994c).

Since the diffusivity of each chemical species is considered as a function of porosity, a common phenomenological relation can be used for describing this function (Bear 1972, Chadam et al. 1986).

$$D_i(\phi) = D_{0i} \phi^q \qquad \left(\frac{3}{2} \leq q \leq \frac{5}{2}\right), \tag{5.4}$$

where D_{0i} is the diffusivity of chemical species i in pure water.

To consider the permeability change caused by a change in porosity, an equation is needed to express the relationship between permeability and porosity. In this regard, Detournay and Cheng (1993) state that "The intrinsic permeability k is generally a function of the pore geometry. In particular, it is strongly dependent on porosity ϕ. According to the Carman-Kozeny law (Scheidegger 1974) which is based on the conceptual model of packing of spheres, a power law relation of $k \propto \phi^3/(1-\phi)^2$ exists. Other models based on different pore geometry give similar power laws. Actual measurements on rocks, however, often yield power law relations with exponents for ϕ significantly larger than 3." Also, Nield and Bejan

(1992) state that "The Carman-Kozeny law is widely used since it seems to be the best simple expression available." For these reasons, the Carman-Kozeny law will be used to calculate permeability k, for a given porosity ϕ.

$$k(\phi) = \frac{k_0(1-\phi_0)^2 \phi^3}{\phi_0^3 (1-\phi)^2}, \tag{5.5}$$

where ϕ_0 and k_0 are the initial reference porosity and permeability of the porous medium respectively.

The source/sink term of chemical species i due to the dissolution/precipitation of solid minerals within the system can be determined in the following manner (Chadam et al. 1986). At the particle level, it is assumed that the average volume of soluble grains is $\overline{V}_p$ and that the density of the soluble grains is D_p, which is defined as the number of the soluble gains per unit medium volume. If the volume fraction of insoluble gains is denoted by $\phi_{insoluble}$, then the final (i.e. maximum) porosity of the porous medium can be denoted by $\phi_f = 1 - \phi_{insoluble}$. In this case, the average volume of soluble grains can be expressed as follows:

$$\overline{V}_p = \frac{\phi_f - \phi}{D_p}, \tag{5.6}$$

At the particle level, the rate of grain-volume change due to a chemical reaction is denoted by R_p, so that the rate of porosity change can be expressed as:

$$\frac{\partial \phi}{\partial t} = -D_p R_p. \tag{5.7}$$

Without loss of generality, it is assumed that the solid grains are dissolved according to the following formula:

$$Solid \Rightarrow \sum_{i=1}^{N} \chi_i X_i, \tag{5.8}$$

where χ_i is the stoichiometric coefficient of the *ith* chemical species; X_i represents chemical species i in the pore-fluid.

It is commonly assumed that the rate of grain-volume change due to a chemical reaction can be expressed as follows (Chadam et al. 1986):

$$R_p = k_{chemical} A_p \left(\prod_{i=1}^{N} C_i^{\chi_i} - K_{eq} \right) \tag{5.9}$$

where A_p is the averaged surface area of soluble grains; $k_{chemical}$ and K_{eq} are the rate constant and equilibrium constant of the chemical reaction respectively.

5.1 Mathematical Background of Chemical Dissolution Front Instability Problems

If the molar density (i.e. moles per volume) of the soluble grains is denoted by ρ_s, then the source/sink term of chemical species i due to the dissolution/precipitation of solid minerals within the system can be expressed as follows:

$$R_i = -\chi_i \rho_s k_{chemical} D_p A_p \left(\prod_{i=1}^{N} C_i^{\chi_i} - K_{eq} \right)$$

$$= -\chi_i \rho_s k_{chemical} \frac{A_p}{V_p}(\phi_f - \phi) \left(\prod_{i=1}^{N} C_i^{\chi_i} - K_{eq} \right). \qquad (5.10)$$

5.1.2 A Particular Case of Reactive Single-Chemical-Species Transport with Consideration of Porosity/Permeability Feedack

If the pore-fluid is incompressible, the governing equations of a reactive single-chemical-species transport problem in a fluid-saturated porous medium can be written as follows:

$$\frac{\partial \phi}{\partial t} - \nabla \bullet [\psi(\phi)\nabla p] = 0, \qquad (5.11)$$

$$\frac{\partial}{\partial t}(\phi C) - \nabla \bullet [\phi D(\phi)\nabla C + C\psi(\phi)\nabla p] + \rho_s k_{chemical} \frac{A_p}{V_p}(\phi_f - \phi)(C - C_{eq}) = 0, \qquad (5.12)$$

$$\frac{\partial \phi}{\partial t} + k_{chemical} \frac{A_p}{V_p}(\phi_f - \phi)(C - C_{eq}) = 0, \qquad (5.13)$$

$$\psi(\phi) = \frac{k(\phi)}{\mu}, \qquad (5.14)$$

where C and C_{eq} are the concentration and equilibrium concentration of the single chemical species. Other quantities in Eqs. (5.11), (5.12), (5.13) and (5.14) are of the same meanings as those defined in Eqs. (5.1), (5.2) (5.3), and (5.9).

Note that Eqs. (5.11) and (5.12) can be derived by substituting the linear average velocity into Eqs. (5.1) and (5.3) with consideration of a single-chemical species.

It needs to be pointed out that for this single-chemical-species system, it is very difficult, even if not impossible, to obtain a complete set of analytical solutions for the pore-fluid pressure, chemical species concentration and porosity within the fluid-saturated porous medium. However, in some special cases, it is possible to obtain analytical solutions for some variables involved in this single-chemical-species system. The first special case to be considered is a problem, in which a planar dissolution front propagates in the full space. Since the dissolution front is a plane, the problem described in Eqs. (5.11), (5.12) and (5.13) degenerates into a one-dimensional problem. For this particular case, analytical solutions can be obtained for both the propagation speed of the dissolution front and the downstream pressure gradient of the pore-fluid. The second special case to be considered is an asymptotic problem, in which the solid molar density greatly exceeds the equilibrium concentration of the chemical species, implying that the region of a considerable porosity change propagates very slowly within the fluid-saturated porous medium. In this particular case, it is possible to derive a complete set of analytical solutions for the pore-fluid pressure, chemical species concentration and porosity within the fluid-saturated porous medium. In addition, it is also possible to investigate the reactive infiltration instability associated with the dissolution front propagation in this particular case (Chadam et al. 1986).

5.1.2.1 The First Special Case

In this special case, the planar dissolution front is assumed to propagate in the positive x direction, so that all quantities are independent of the transverse coordinates y and z. For this reason, Eqs. (5.11), (5.12) and (5.13) can be rewritten as follows:

$$\frac{\partial \phi}{\partial t} - \frac{\partial}{\partial x}\left[\psi(\phi)\frac{\partial p}{\partial x}\right] = 0, \tag{5.15}$$

$$\frac{\partial}{\partial t}(\phi C) - \frac{\partial}{\partial x}\left[\phi D(\phi)\frac{\partial C}{\partial x} + C\psi(\phi)\frac{\partial p}{\partial x}\right] + \rho_s k_{chemical}\frac{A_p}{V_p}(\phi_f - \phi)(C - C_{eq}) = 0, \tag{5.16}$$

$$\frac{\partial \phi}{\partial t} + k_{chemical}\frac{A_p}{V_p}(\phi_f - \phi)(C - C_{eq}) = 0. \tag{5.17}$$

If the chemical species is initially in an equilibrium state and fresh pore-fluid is injected at the location of x approaching negative infinite, then the boundary conditions of this special problem are expressed as

$$\lim_{x \to -\infty} C = 0, \quad \lim_{x \to -\infty} \phi = \phi_f, \quad \lim_{x \to -\infty} \frac{\partial p}{\partial x} = p'_{fx} \quad \text{(Upstream boundary)}, \tag{5.18}$$

$$\lim_{x \to \infty} C = C_{eq}, \quad \lim_{x \to \infty} \phi = \phi_0, \quad \lim_{x \to \infty} \frac{\partial p}{\partial x} = p'_{0x} \quad \text{(Downstream boundary)}, \tag{5.19}$$

where ϕ_0 is the initial porosity of the porous medium; p'_{fx} is the pore-fluid pressure gradient as x approaching negative infinite in the upstream of the pore-fluid flow; p'_{0x} is the unknown pore-fluid pressure gradient as x approaching positive infinite in the downstream of the pore-fluid flow. Since p'_{fx} drives the pore-fluid flow continuously

5.1 Mathematical Background of Chemical Dissolution Front Instability Problems

along the positive x direction, it has a negative algebraic value (i.e. $p'_{fx} < 0$) in this analysis.

If the propagation speed of the planar dissolution front is denoted by v_{front}, then it is possible to transform a moving boundary problem of the dissolution front (in an x-t coordinate system) into a steady-state boundary problem of the dissolution front (in an $\xi - t$ coordinate system) using the following coordinate mapping:

$$\xi = x - v_{front} t. \tag{5.20}$$

It is necessary to relate partial derivatives with respect to ξ and t to those with respect to x and t (Turcotte and Schubert 1982).

$$\left(\frac{\partial}{\partial t}\right)_\xi = \left(\frac{\partial}{\partial t}\right)_x + \frac{\partial}{\partial x}\frac{\partial x}{\partial t} = \left(\frac{\partial}{\partial t}\right)_x + v_{front}\frac{\partial}{\partial x}, \tag{5.21}$$

$$\left(\frac{\partial}{\partial \xi}\right)_t = \left(\frac{\partial}{\partial x}\right)_t, \tag{5.22}$$

where derivatives are taken with x or t held constant as appropriate.

Since the transformed system in the $\xi - t$ coordinate system is in a steady state, the following equations can be derived from Eqs. (5.21) and (5.22).

$$\left(\frac{\partial}{\partial t}\right)_x = -v_{front}\frac{\partial}{\partial \xi}, \tag{5.23}$$

$$\left(\frac{\partial}{\partial \xi}\right)_t = \left(\frac{\partial}{\partial x}\right)_t. \tag{5.24}$$

Substituting Eqs. (5.23) and (5.24) into Eqs. (5.15), (5.16) and (5.17) yields the following equations:

$$\frac{\partial}{\partial \xi}\left[\psi(\phi)\frac{\partial p}{\partial \xi} + v_{front}\phi\right] = 0, \tag{5.25}$$

$$\frac{\partial}{\partial \xi}\left[\phi D(\phi)\frac{\partial C}{\partial \xi} + C\psi(\phi)\frac{\partial p}{\partial \xi} + v_{front}(C - \rho_s)\phi\right] = 0, \tag{5.26}$$

$$v_{front}\frac{\partial \phi}{\partial \xi} - k_{chemical}\frac{A_p}{V_p}(\phi_f - \phi)(C - C_{eq}) = 0. \tag{5.27}$$

Integrating Eqs. (5.25) and (5.26) from negative infinite to positive infinite and using the boundary conditions (i.e. Eqs. (5.18) and (5.19)) yields the following equations:

$$C_{eq}\psi(\phi_0)p'_{ox} + v_{front}\phi_0(C_{eq} - \rho_s) + v_{front}\phi_f\rho_s = 0, \qquad (5.28)$$

$$\psi(\phi_0)p'_{ox} + v_{front}\phi_0 - \psi(\phi_f)p'_{fx} - v_{front}\phi_f = 0. \qquad (5.29)$$

Solving Eqs. (5.28) and (5.29) simultaneously results in the following analytical solutions:

$$v_{front} = \frac{-\psi(\phi_0)p'_{ox}C_{eq}}{\phi_0 C_{eq} + (\phi_f - \phi_0)\rho_s} = \frac{u_{0x}C_{eq}}{\phi_0 C_{eq} + (\phi_f - \phi_0)\rho_s}, \qquad (5.30)$$

$$p'_{ox} = \frac{\psi(\phi_f)[\phi_0 C_{eq} + (\phi_f - \phi_0)\rho_s]}{\psi(\phi_0)[\phi_0 C_{eq} + (\phi_f - \phi_0)(\rho_s + C_{eq})]} p'_{fx}, \qquad (5.31)$$

where u_{0x} is the Darcy velocity in the far downstream of the flow as x approaches positive infinite. Using Darcy's law, u_{0x} can be expressed as

$$u_{0x} = \frac{\phi_0 C_{eq} + (\phi_f - \phi_0)\rho_s}{\phi_0 C_{eq} + (\phi_f - \phi_0)(\rho_s + C_{eq})} u_{fx}, \qquad (5.32)$$

where u_{fx} is the Darcy velocity in the far upstream of the flow as x approaches negative infinite.

If the finite element method is used to solve this special problem, the accuracy of the finite element simulation can be conveniently evaluated by comparing the numerical solutions with the analytical ones for both the propagation speed of the planar dissolution front (i.e. v_{front}) and the Darcy velocity in the far downstream of the flow as x approaches positive infinite (i.e. u_{0x}).

5.1.2.2 The Second Special Case (Base Solutions for a Stable State)

Since the solid molar density greatly exceeds the equilibrium concentration of the chemical species, a small parameter can be defined as follows:

$$\varepsilon = \frac{C_{eq}}{\rho_s} \ll 1. \qquad (5.33)$$

To facilitate the theoretical analysis in the limit case of ε approaching zero, the following dimensionless parameters and variables can be defined.

$$\bar{x} = \frac{x}{L^*}, \qquad \bar{y} = \frac{y}{L^*}, \qquad \bar{z} = \frac{z}{L^*}, \qquad (5.34)$$

$$\bar{C} = \frac{C}{C_{eq}}, \qquad \bar{p} = \frac{p}{p^*}, \qquad \bar{\vec{u}} = \frac{\vec{u}}{u^*}, \qquad (5.35)$$

5.1 Mathematical Background of Chemical Dissolution Front Instability Problems

$$\tau = \frac{t}{t^*}\varepsilon, \tag{5.36}$$

where τ is a slow dimensionless time to describe the slowness of the chemical dissolution that takes place in the system. Other characteristic parameters used in Eqs. (5.34), (5.35) and (5.36) can be expressed as follows:

$$t^* = \frac{\overline{V}_p}{k_{chemical}A_pC_{eq}}, \qquad L^* = \sqrt{\phi_f D(\phi_f)t^*}, \tag{5.37}$$

$$p^* = \frac{\phi_f D(\phi_f)}{\psi(\phi_f)}, \qquad u^* = \frac{\phi_f D(\phi_f)}{L^*}, \tag{5.38}$$

$$D^*(\phi) = \frac{\phi D(\phi)}{\phi_f D(\phi_f)}, \qquad \psi^*(\phi) = \frac{\psi(\phi)}{\psi(\phi_f)}, \tag{5.39}$$

Substituting Eqs. (5.34), (5.35), (5.36), (5.37), (5.38) and (5.39) into Eqs. (5.11), (5.12) and (5.13) yields the following dimensionless equations:

$$\varepsilon\frac{\partial \phi}{\partial \tau} - \nabla \bullet [\psi^*(\phi)\nabla\overline{p}] = 0, \tag{5.40}$$

$$\varepsilon\frac{\partial}{\partial \tau}(\phi\overline{C}) - \nabla \bullet [D^*(\phi)\nabla\overline{C} + \overline{C}\psi^*(\phi)\nabla\overline{p}] - \frac{\partial \phi}{\partial \tau} = 0, \tag{5.41}$$

$$\varepsilon\frac{\partial \phi}{\partial \tau} + (\phi_f - \phi)(\overline{C} - 1) = 0. \tag{5.42}$$

Similarly, the boundary conditions for this special case can be expressed in a dimensionless form as follows:

$$\lim_{\overline{x}\to\infty}\overline{C} = 1, \qquad \lim_{\overline{x}\to\infty}\phi = \phi_0, \qquad \lim_{\overline{x}\to\infty}\frac{\partial \overline{p}}{\partial \overline{x}} = \overline{p}'_{0x} \quad \text{(downstream boundary)}, \tag{5.43}$$

$$\lim_{\overline{x}\to-\infty}\overline{C} = 0, \qquad \lim_{\overline{x}\to-\infty}\phi = \phi_f, \qquad \lim_{\overline{x}\to-\infty}\frac{\partial \overline{p}}{\partial \overline{x}} = \overline{p}'_{fx} \quad \text{(upstream boundary)}. \tag{5.44}$$

It is noted that the propagation front due to chemical dissolution divides the problem domain into two regions, an upstream region and a downstream region, relative to the propagation front. Across this propagation front, the porosity undergoes a jump from its initial value into its final value. Thus, this dissolution-front propagation problem can be considered as a Stefan moving boundary problem (Chadam et al. 1986). In the limit case of ε approaching zero, the corresponding governing equations for the dimensionless variables of the problem in both the downstream region and the upstream region can be expressed below:

$$\overline{C} = 1, \qquad \nabla^2 \overline{p} = 0, \qquad \phi = \phi_0 \qquad \text{(in the downstream region)}, \qquad (5.45)$$

$$\nabla \cdot (\nabla \overline{C} + \overline{C} \nabla \overline{p}) = 0, \qquad \nabla^2 \overline{p} = 0, \qquad \phi = \phi_f \qquad \text{(in the upstream region)}. \qquad (5.46)$$

If the chemical dissolution front is denoted by $S(\overline{x}, \tau) = 0$, then the dimensionless pressure, chemical species concentration and mass fluxes of both the chemical species and the pore-fluid should be continuous on $S(\overline{x}, \tau) = 0$. This leads to the following interface conditions for this moving-front problem:

$$\lim_{S \to 0^-} \overline{C} = \lim_{S \to 0^+} \overline{C}, \qquad \lim_{S \to 0^-} \overline{p} = \lim_{S \to 0^+} \overline{p}, \qquad (5.47)$$

$$\lim_{S \to 0^-} \frac{\partial \overline{C}}{\partial n} = \overline{v}_{front}(\phi_f - \phi_0), \qquad \lim_{S \to 0^-} \frac{\partial \overline{p}}{\partial n} = \frac{\psi(\phi_0)}{\psi(\phi_f)} \lim_{S \to 0^+} \frac{\partial \overline{p}}{\partial n}, \qquad (5.48)$$

where n is the unit normal vector of the moving dissolution front.

When the planar dissolution front is under stable conditions, the base solutions for this special problem can be derived from Eqs. (5.45) and (5.46) with the related boundary and interface conditions (i.e. Eqs. (5.43), (5.44), (5.47) and (5.48)). The resulting base solutions are expressed as follows:

$$\overline{C}(\xi) = 1, \qquad \overline{p}(\xi) = \overline{p}'_{0x}\xi + \overline{p}_{C1}, \qquad \phi = \phi_0 \qquad \text{(in the downstream region)}, \qquad (5.49)$$

$$\overline{C}(\xi) = \exp(-\overline{p}'_{fx}\xi), \qquad \overline{p}(\xi) = \overline{p}'_{fx}\xi + \overline{p}_{C2}, \qquad \phi = \phi_f \qquad \text{(in the upstream region)}, \qquad (5.50)$$

where $\overline{p}_{C1}$ and $\overline{p}_{C2}$ are two constants to be determined. For example, $\overline{p}_{C1}$ can be determined by setting the dimensionless pressure $\overline{p}(\xi)$ to be a constant at a prescribed location of the downstream region, while $\overline{p}_{C2}$ can be determined using the pressure continuity condition at the interface between the upstream and downstream regions. Other parameters are defined below:

$$\xi = \overline{x} - \overline{v}_{front}\tau, \qquad \overline{p}'_{0x} = \frac{\psi(\phi_f)}{\psi(\phi_0)}\overline{p}'_{fx}, \qquad \overline{v}_{front} = -\frac{\overline{p}'_{fx}}{\phi_f - \phi_0}. \qquad (5.51)$$

Therefore, if the finite element method is used to solve the second special problem, the accuracy of the finite element simulation can be conveniently evaluated by comparing the numerical solutions with a complete set of analytical solutions including porosity, the location of the chemical dissolution front, the dimensionless chemical-species concentration and the dimensionless pressure.

5.1.2.3 The Second Special Case (Perturbation Solutions for an Unstable State)

When a reactive transport system represented by the above-mentioned second special problem is stable, the planar dissolution front remains planar, even though both small perturbations of the dissolution front and the feedback effect of porosity/permeability change are simultaneously considered in the analysis. However, when the reactive transport system is unstable, the planar dissolution front can change from a planar shape into a complicated one. The instability of the above-mentioned second special problem can be determined using a linear stability analysis (Chadam et al. 1986, 1988, Ortoleva et al. 1987). The main purpose of conducting such a linear stability analysis is to determine the critical condition under which the chemical dissolution front of the reactive transport system becomes unstable.

If a small time-dependent perturbation is added to the planar dissolution front, then the total solution of the system is equal to the summation of the base solution and the perturbed solution of the system.

$$S(\xi, \bar{y}, \tau) = \xi - \delta \exp(\omega \tau) \cos(m\bar{y}), \tag{5.52}$$

$$\bar{P}_{total}(\xi, \bar{y}, \tau) = \bar{p}(\xi, \tau) + \delta \hat{p}(\xi) \exp(\omega \tau) \cos(m\bar{y}), \tag{5.53}$$

$$\bar{C}_{total}(\xi, \bar{y}, \tau) = \bar{C}(\xi, \tau) + \delta \hat{C}(\xi) \exp(\omega \tau) \cos(m\bar{y}), \tag{5.54}$$

where ω is the growth rate of the perturbation; m is the wavenumber of the perturbation; δ is the amplitude of the perturbation and $\delta \ll 1$ by the definition of a linear stability analysis.

Since $S(\xi, \bar{y}, \tau)$ is a function of coordinates ξ and $\bar{y}$, the following derivatives exist mathematically:

$$\left(\frac{\partial}{\partial \xi}\right)_\xi = \frac{\partial S}{\partial \xi} \frac{\partial}{\partial S} = \left(\frac{\partial}{\partial \xi}\right)_S, \tag{5.55}$$

$$\left(\frac{\partial}{\partial \bar{y}}\right)_\xi = \frac{\partial S}{\partial \bar{y}} \frac{\partial}{\partial S} + \left(\frac{\partial}{\partial \bar{y}}\right)_S = \frac{\partial S}{\partial \bar{y}} \left(\frac{\partial}{\partial \xi}\right)_S + \left(\frac{\partial}{\partial \bar{y}}\right)_S, \tag{5.56}$$

$$\left(\frac{\partial^2}{\partial \xi^2}\right)_\xi = \left(\frac{\partial^2}{\partial \xi^2}\right)_S, \tag{5.57}$$

$$\left(\frac{\partial^2}{\partial \bar{y}^2}\right)_\xi = \frac{\partial^2 S}{\partial \bar{y}^2} \frac{\partial}{\partial \xi} + \left(\frac{\partial S}{\partial \bar{y}}\right)^2 \frac{\partial^2}{\partial \xi^2} + 2\frac{\partial S}{\partial \bar{y}} \frac{\partial^2}{\partial \xi \partial \bar{y}} + \left(\frac{\partial^2}{\partial \bar{y}^2}\right)_S. \tag{5.58}$$

It is noted that the total solutions expressed in Eqs. (5.53) and (5.54) must satisfy the governing equations that are expressed in Eqs. (5.45) and (5.46). With consideration of Eq. (5.58), the first-order perturbation equations of this system can be expressed as

$$\hat{C} = 0, \qquad \frac{\partial^2 \hat{p}}{\partial \xi^2} - m^2 \hat{p} + m^2 \overline{p}'_{0x} = 0 \qquad \text{(in the downstream region)}, \quad (5.59)$$

$$\frac{\partial^2 \hat{C}}{\partial \xi^2} + \overline{p}'_{fx} \frac{\partial \hat{C}}{\partial \xi} - m^2 \hat{C} - m^2 \overline{p}'_{fx} \exp(-\overline{p}'_{fx}\xi) - \overline{p}'_{fx} \exp(-\overline{p}'_{fx}\xi) \frac{\partial \hat{p}}{\partial \xi} = 0,$$

$$\frac{\partial^2 \hat{p}}{\partial \xi^2} - m^2 \hat{p} + m^2 \overline{p}'_{fx} = 0 \qquad \text{(in the upstream region)}. \quad (5.60)$$

The corresponding boundary conditions of the first-order perturbation problem are:

$$\hat{C} = 0, \qquad \lim_{x \to \infty} \frac{\partial \hat{p}}{\partial \xi} = 0 \qquad \text{(downstream boundary)}, \quad (5.61)$$

$$\lim_{x \to -\infty} \hat{C} = 0, \qquad \lim_{x \to -\infty} \frac{\partial \hat{p}}{\partial \xi} = 0 \qquad \text{(upstream boundary)}. \quad (5.62)$$

Similarly, the interface conditions for this first-order perturbation problem can be expressed as follows:

$$\hat{C} = 0, \qquad \lim_{S \to 0^-} \hat{p} = \lim_{S \to 0^+} \hat{p}, \quad (5.63)$$

$$\lim_{S \to 0^-} \frac{\partial \hat{C}}{\partial n} = \omega(\phi_f - \phi_0), \qquad \lim_{S \to 0^-} \frac{\partial \hat{p}}{\partial n} = \frac{\psi(\phi_0)}{\psi(\phi_f)} \lim_{S \to 0^+} \frac{\partial \hat{p}}{\partial n}. \quad (5.64)$$

Solving Eqs. (5.59) and (5.60) with the boundary and interface conditions (i.e. Eqs. (5.61) and (5.62)) yields the following analytical results:

$$\hat{C} = 0, \qquad \hat{p}(\xi) = \overline{p}'_{0x} \left[1 - \frac{1-\beta}{1+\beta} \exp(-|m|\xi) \right] \qquad \text{(in the downstream region)}, \quad (5.65)$$

$$\hat{C}(\xi) = -\overline{p}'_{fx} \left\{ \exp(-\overline{p}'_{fx}\xi) - \frac{2}{1+\beta} \exp(\sigma \xi) + \frac{1-\beta}{1+\beta} \exp[(|m| - \overline{p}'_{fx})\xi] \right\},$$

$$\hat{p}(\xi) = \overline{p}'_{fx} \left[1 + \frac{1-\beta}{1+\beta} \exp(|m|\xi) \right] \qquad \text{(in the upstream region)}, \quad (5.66)$$

5.1 Mathematical Background of Chemical Dissolution Front Instability Problems

where

$$\beta = \frac{\psi(\phi_0)}{\psi(\phi_f)} = \frac{k(\phi_0)}{k(\phi_f)}, \quad (5.67)$$

$$\sigma = \frac{\sqrt{(\overline{p}'_{fx})^2 + 4m^2} - \overline{p}'_{fx}}{2}. \quad (5.68)$$

Substituting Eq. (5.66) into Eq. (5.64) yields the following equation for the growth rate of the small perturbation:

$$\omega(m) = \frac{-\overline{p}'_{fx}}{(1+\beta)(\phi_f - \phi_0)}[-\overline{p}'_{fx} - \sqrt{(\overline{p}'_{fx})^2 + 4m^2} + (1-\beta)|m|]. \quad (5.69)$$

Equation (5.69) clearly indicates that the planar dissolution front of the reactive transport system, which is described by the above-mentioned second special problem, is stable to short wavelength (i.e. large wavenumber m) perturbations but it is unstable to long wavelength (i.e. small wavenumber m) perturbations.

Letting $\omega(m) = 0$ yields the following critical condition, under which the reactive transport system can become unstable.

$$\overline{p}'_{fx}\big|_{critical} = -\frac{(3-\beta)(1+\beta)}{2(1-\beta)}, \quad (5.70)$$

where $\overline{p}'_{fx}\big|_{critical}$ is the critical value of the generalized dimensionless pressure gradient in the far upstream direction as x approaching negative infinite (Zhao et al. 2008e). Since $\overline{p}'_{fx}\big|_{critical}$ is usually of a negative value, the following critical Zhao number is defined to judge the instability of the reactive transport system:

$$Zh_{critical} = -\overline{p}'_{fx}\big|_{critical} = \frac{(3-\beta)(1+\beta)}{2(1-\beta)}. \quad (5.71)$$

Thus, the Zhao number of the reactive transport system can be defined as follows:

$$Zh = -\overline{P}'_{fx} = -\frac{p'_{fx} L^*}{p^*} = -\frac{k(\phi_f) L^* p'_{fx}}{\phi_f \mu D(\phi_f)} = \frac{v_{flow}}{\sqrt{\phi_f D(\phi_f)}} \sqrt{\frac{\overline{V}_p}{k_{chemical} A_p C_{eq}}}. \quad (5.72)$$

Using Eqs. (5.71) and (5.72), a criterion can be established to judge the instability of a chemical dissolution front associated with the particular chemical system in this investigation. If $Zh > Zh_{critical}$, then the chemical dissolution front of the reactive transport system becomes unstable, while if $Zh < Zh_{critical}$, then the chemical dissolution front of the reactive transport system is stable. The case of $Zh = Zh_{critical}$ represents a situation where the chemical dissolution front of the reactive transport system is neutrally unstable, implying that the introduced small perturbation can be maintained but it does not grow in the corresponding reactive transport system.

Clearly, Eq. (5.72) indicates that for the reactive chemical-species transport considered in this investigation, the dissolution-enhanced permeability destabilizes the instability of the chemical dissolution front, while the dissolution-enhanced diffusivity stabilizes the instability of the chemical dissolution front. If the shape factor of soluble grains is represented by $\theta = \overline{V}_p / \overline{A}_p$, then an increase in the shape factor of soluble grains can destabilize the instability of the chemical dissolution front, indicating that the instability likelihood of a porous medium comprised of irregular grains, is higher than that of a porous medium comprised of regular spherical grains. Similarly, an increase in either the equilibrium concentration of the chemical species or the chemical reaction constant of the dissolution reaction can cause the stabilization of the chemical dissolution front, for the reactive chemical-species transport considered in this investigation.

To understand the physical meanings of each term in the Zhao number, Equation (5.72) can be rewritten in the following form:

$$Zh = F_{Advection} F_{Diffusion} F_{chemical} F_{Shape}, \qquad (5.73)$$

where $F_{Advection}$ is a term to represent the solute advection; $F_{Diffusion}$ is a term to represent the solute diffusion/dispersion; $F_{chemical}$ is a term to represent the chemical kinetics of the dissolution reaction; F_{Shape} is a term to represent the shape factor of the soluble mineral in the fluid-rock interaction system. These terms can be expressed as follows:

$$F_{Advection} = v_{flow}, \qquad (5.74)$$

$$F_{Diffusion} = \frac{1}{\sqrt{\phi_f D(\phi_f)}}, \qquad (5.75)$$

$$F_{chemical} = \sqrt{\frac{1}{k_{chemical} C_{eq}}}, \qquad (5.76)$$

$$F_{Shape} = \sqrt{\frac{\overline{V}_p}{\overline{A}_p}}. \qquad (5.77)$$

Equations (5.73), (5.74), (5.75), (5.75) and (5.77) clearly indicate that the Zhao number is a dimensionless number that can be used to represent the geometrical, hydrodynamic, thermodynamic and chemical kinetic characteristics of a fluid-rock system in a comprehensive manner. This dimensionless number reveals the intimate interaction between solute advection, solution diffusion/dispersion, chemical kinetics and mineral geometry in a reactive transport system.

5.2 Proposed Segregated Algorithm for Simulating the Morphological Evolution of a Chemical Dissolution Front

Although analytical solutions can be obtained for the above-mentioned special cases, it is very difficult, if not impossible, to predict analytically the complicated morphological evolution process of a planar dissolution front in the case of the chemical dissolution system becoming supercritical. As an alternative, numerical methods are suitable to overcome this difficulty. Since numerical methods are approximate solution methods, they must be verified before they are used to solve any new type of scientific and engineering problem. For this reason, the main purpose of this section is to propose a numerical procedure for simulating how a planar dissolution front evolves into a complicated morphological front. To verify the accuracy of the numerical solution, a benchmark problem is constructed from the theoretical analysis in the previous section. As a result, the numerical solution obtained from the benchmark problem can be compared with the corresponding analytical solution. After the proposed numerical procedure is verified, it will be used to simulate the complicated morphological evolution process of a planar dissolution front in the case of the chemical dissolution system becoming supercritical.

5.2.1 Formulation of the Segregated Algorithm for Simulating the Evolution of Chemical Dissolution Fronts

In this section, Eqs. (5.40), (5.41) and (5.42) are solved using the proposed numerical procedure, which is a combination of both the finite element method and the finite difference method. The finite element method is used to discretize the geometrical shape of the problem domain, while the finite difference method is used to discretize the dimensionless time. Since the system described by these equations is highly nonlinear, the segregated algorithm, in which Eqs. (5.40), (5.41) and (5.42) are solved separately in a sequential manner, is used to derive the formulation of the proposed numerical procedure.

For a given dimensionless time-step, $\tau + \Delta\tau$, the porosity can be denoted by $\phi_{\tau+\Delta\tau} = \phi_\tau + \Delta\phi_{\tau+\Delta\tau}$, where ϕ_τ is the porosity at the previous time-step and $\Delta\phi_{\tau+\Delta\tau}$ is the porosity increment at the current time-step. Using the backward difference scheme, Eq. (5.42) can be written as follows:

$$\left[\frac{\varepsilon}{\Delta\tau} + (1 - \overline{C}_{\tau+\Delta\tau})\right] \Delta\phi_{\tau+\Delta\tau} = (\phi_f - \phi_\tau)(1 - \overline{C}_{\tau+\Delta\tau}), \quad (5.78)$$

where $\overline{C}_{\tau+\Delta\tau}$ is the dimensionless concentration at the current time-step; $\Delta\tau$ is the dimensionless time increment at the current time-step.

Mathematically, there exist the following relationships in the finite difference sense:

$$\varepsilon\frac{\partial(\phi\overline{C})}{\partial\tau} = \varepsilon\frac{\Delta(\phi_{\tau+\Delta\tau}\overline{C}_{\tau+\Delta\tau})}{\Delta\tau} = \varepsilon\overline{C}_{\tau+\Delta\tau}\frac{\Delta\phi_{\tau+\Delta\tau}}{\Delta\tau} + \varepsilon\phi_{\tau+\Delta\tau}\frac{\Delta(\overline{C}_{\tau+\Delta\tau})}{\Delta\tau}, \tag{5.79}$$

$$\varepsilon\frac{\partial\phi}{\partial\tau} = \varepsilon\frac{\Delta(\phi_{\tau+\Delta\tau})}{\Delta\tau} = (1-\overline{C}_{\tau+\Delta\tau})(\phi_f - \phi_{\tau+\Delta\tau}), \tag{5.80}$$

$$\nabla\bullet[D^*(\phi)\nabla\overline{C}] = \nabla\bullet[D^*(\phi_{\tau+\Delta\tau})\nabla\overline{C}_{\tau+\Delta\tau}], \tag{5.81}$$

$$\begin{aligned}\nabla\bullet[\overline{C}\psi^*(\phi)\nabla\overline{p}] &= \overline{C}\nabla\bullet[\psi^*(\phi)\nabla\overline{p}] + \nabla\overline{p}\bullet[\psi^*(\phi)\nabla\overline{C}]\\ &= \overline{C}_{\tau+\Delta\tau}\nabla\bullet[\psi^*(\phi_{\tau+\Delta\tau})\nabla\overline{p}_{\tau+\Delta\tau}]\\ &\quad + \nabla\overline{p}_{\tau+\Delta\tau}\bullet[\psi^*(\phi_{\tau+\Delta\tau})\nabla\overline{C}_{\tau+\Delta\tau}]\end{aligned} \tag{5.82}$$

Substituting Eqs. (5.79), (5.80), (5.81) and (5.82) into Eq. (5.41) yields the following finite difference equation:

$$\begin{aligned}\left[\frac{\varepsilon}{\Delta\tau}\phi_{\tau+\Delta\tau} + \frac{1}{\varepsilon}(\phi_f - \phi_{\tau+\Delta\tau})\right]\overline{C}_{\tau+\Delta\tau} &- \nabla\bullet[D^*(\phi_{\tau+\Delta\tau})\nabla\overline{C}_{\tau+\Delta\tau}]\\ &- \nabla\overline{p}_{\tau+\Delta\tau}\bullet[\psi^*(\phi_{\tau+\Delta\tau})\nabla\overline{C}_{\tau+\Delta\tau}]\\ &= \frac{\varepsilon}{\Delta\tau}\phi_{\tau+\Delta\tau}\overline{C}_\tau + \frac{1}{\varepsilon}(\phi_f - \phi_{\tau+\Delta\tau})\end{aligned} \tag{5.83}$$

Similarly, Eq. (5.40) can be rewritten in the following discretized form:

$$\nabla\bullet[\psi^*(\phi)\nabla\overline{p}] = \nabla\bullet[\psi^*(\phi_{\tau+\Delta\tau})\nabla\overline{p}_{\tau+\Delta\tau}] = (1-\overline{C}_{\tau+\Delta\tau})(\phi_f - \phi_{\tau+\Delta\tau}). \tag{5.84}$$

Using the proposed segregated scheme and finite element method, Eqs. (5.78), (5.83) and (5.84) are solved separately and sequentially for the porosity, dimensionless concentration and dimensionless pressure at the current time-step. Note that when Eq. (5.78) is solved using the finite element method, the dimensionless concentration at the current time-step is not known. Similarly, when Eq. (5.83) is solved using the finite element method, the dimensionless pressure at the current time-step remains unknown. This indicates that these three equations are fully coupled so that an iteration scheme needs to be used to solve them sequentially. At the first iteration step, the dimensionless concentration at the previous time-step is used as a reasonable guess of the dimensionless concentration at the current time-step when Eq. (5.78) is solved for the porosity. In a similar way, the dimensionless pressure at the previous time-step is used as a reasonable guess for the current time-step when Eq. (5.83) is solved for the dimensionless concentration. The resulting approximate porosity and dimensionless concentration can be used when Eq. (5.84) is solved for the dimensionless pressure. At the second iteration step, the same procedure as used in the first iteration step is followed, so that the following convergence criterion can be established after the second iteration step.

5.2 Segregated Algorithm for Simulating the Morphological Evolution

$$E = Max \left(\sqrt{\sum_{i=1}^{N_\phi} \left(\phi_{i,\tau+\Delta\tau}^k - \phi_{i,\tau+\Delta\tau}^{k-1}\right)^2}, \sqrt{\sum_{i=1}^{N_{\overline{C}}} \left(\overline{C}_{i,\tau+\Delta\tau}^k - \overline{C}_{i,\tau+\Delta\tau}^{k-1}\right)^2}, \right.$$
$$\left. \sqrt{\sum_{i=1}^{N_{\overline{p}}} \left(\overline{p}_{i,\tau+\Delta\tau}^k - \overline{p}_{i,\tau+\Delta\tau}^{k-1}\right)^2} \right) < \overline{E},$$
(5.85)

where E and $\overline{E}$ are the maximum error at the k-th iteration step and the allowable error limit; N_ϕ, $N_{\overline{C}}$ and $N_{\overline{p}}$ are the total numbers of the degrees-of-freedom for the porosity, dimensionless concentration and dimensionless pressure respectively; k is the index number at the current iteration step and $k-1$ is the index number at the previous iteration step; $\phi_{i,\tau+\Delta\tau}^k$, $\overline{C}_{i,\tau+\Delta\tau}^k$ and $\overline{p}_{i,\tau+\Delta\tau}^k$ are the porosity, dimensionless concentration and dimensionless pressure of node i at both the current time-step and the current iteration step; $\phi_{i,\tau+\Delta\tau}^{k-1}$, $\overline{C}_{i,\tau+\Delta\tau}^{k-1}$ and $\overline{p}_{i,\tau+\Delta\tau}^{k-1}$ are the porosity, dimensionless concentration and dimensionless pressure of node i at the current time-step but at the previous iteration step. It is noted that $k \geq 2$ in Eq. (5.85).

The convergence criterion is checked after the second iteration step. If the convergence criterion is not met, then the iteration is repeated at the current time-step. Otherwise, the convergence solution is obtained at the current time step and the solution procedure goes to the next time-step until the final time-step is reached.

5.2.2 Verification of the Segregated Algorithm for Simulating the Evolution of Chemical Dissolution Fronts

The main and ultimate purpose of a numerical simulation is to provide numerical solutions for practical problems in a real world. These practical problems are impossible and impractical to solve analytically. Since numerical methods are the basic foundation of a numerical simulation, only an approximate solution can be obtained from a computational model, which is the discretized description of a continuum mathematical model. Due to inevitable round-off errors in computation and discretized errors in temporal and spatial variables, it is necessary to verify the proposed numerical procedure so that meaningful numerical results can be obtained from a discretized computational model. For this reason, a benchmark problem, for which the analytical solutions are available, is considered in this section.

Figure 5.1 shows the geometry and boundary conditions of the coupled problem between porosity, pore-fluid pressure and reactive chemical-species transport within a fluid-saturated porous medium. For this benchmark problem, the dimensionless-pressure gradient (i.e. $\overline{p}_{fx}' = -1$) is applied on the left boundary, implying that there is a horizontal throughflow from the left to the right of the computational model. In this case, the Zhao number of the reactive transport system is unity. The dimensionless height and width of the computational model are 5 and 10 respectively. Except for the left boundary, the initial porosity of the porous medium is 0.1, while the

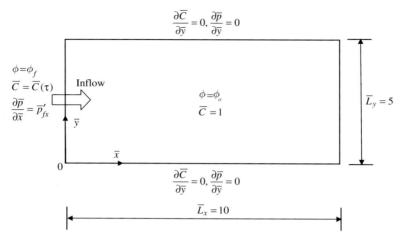

Fig. 5.1 Geometry and boundary conditions of the reactive infiltration problem

initial dimensionless-concentration is one within the computational domain. The final porosity after depletion of the soluble mineral is 0.2. This final porosity is applied on the left boundary as a boundary condition of the computational domain. The permeability of the porous medium is calculated using the Carman-Kozeny formula, which has the power of 3 in the power law. The diffusivity of chemical species is calculated using the power law, which has the power of 2. Both the top and the bottom boundaries are assumed to be impermeable for the pore-fluid and chemical species. The ratio of the equilibrium concentration to the solid molar density of the chemical species is assumed to be 0.01, while the dimensionless time-step is set to be 0.005 in the computation. Since the computational domain of the benchmark problem is of finite size, a time-dependent-dimensionless-concentration boundary condition (i.e. $\overline{C}(\tau) = \exp(\overline{p}'_{fx}\overline{v}_{front}\tau)$) needs to be applied on the left boundary so that the numerical solutions can be compared with the analytical solutions derived in the previous section. Using the above-mentioned parameters, the critical Zhao number of the system is approximately equal to 1.77. Since the Zhao number of the system is smaller than its critical value, the coupled system considered in this section is sub-critical so that a planar dissolution front remains planar during its propagation within the system. The dimensionless speed of the dissolution front propagation is equal to 10, which is determined using Eq. (5.51). To simulate appropriately the propagation of the dissolution front, the whole computational domain is simulated by 19701 four-node rectangular elements of 20000 nodal points in total.

Figures 5.2, 5.3 and 5.4 show the comparison of numerical solutions with analytical ones for the porosity, dimensionless concentration and dimensionless pressure distributions within the computational domain at three different time instants. In these figures, the thick line shows the numerical results, while the thin line shows the corresponding analytical solutions, which can be determined from Eqs. (5.49)

5.2 Segregated Algorithm for Simulating the Morphological Evolution

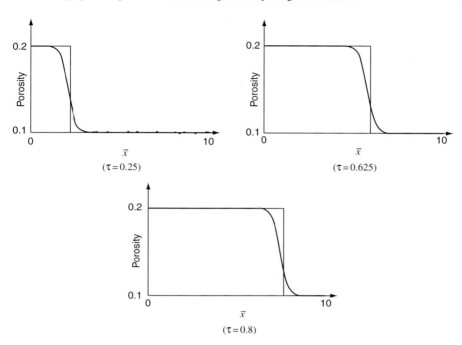

Fig. 5.2 Comparison of numerical solutions with analytical ones at different time instants (Porosity): the *thick line* shows the numerical results, while the *thin line* shows the corresponding analytical solutions

and (5.50) with the boundary condition of $\overline{p}(\overline{L}_x, \tau) = 100$ at the right boundary of the computational model. The resulting analytical solutions are expressed as follows:

$$\overline{C}(\overline{x}, \tau) = 1, \qquad \phi(\overline{x}, \tau) = \phi_0 \qquad (\overline{x} > \overline{v}_{front}\tau), \tag{5.86}$$

$$\overline{p}(\overline{x}, \tau) = -\overline{p}'_{0x}(\overline{L}_x - \overline{x}) + 100 \qquad (\overline{x} > \overline{v}_{front}\tau), \tag{5.87}$$

$$\overline{C}(\overline{x}, \tau) = \exp[-\overline{p}'_{fx}(\overline{x} - \overline{v}_{front}\tau)], \qquad \phi(\overline{x}, \tau) = \phi_f \qquad (\overline{x} \leq \overline{v}_{front}\tau), \tag{5.88}$$

$$\overline{p}(\overline{x}, \tau) = \overline{p}'_{fx}(\overline{x} - \overline{v}_{front}\tau) - \overline{p}'_{0x}(\overline{L}_x - \overline{v}_{front}\tau) + 100 \qquad (\overline{x} \leq \overline{v}_{front}\tau). \tag{5.89}$$

From these results, it can be observed that the numerical solutions agree very well with the analytical solutions, indicating that the proposed numerical procedure is capable of simulating the planar dissolution-front propagation within the fluid-saturated porous medium. As expected, the porosity propagation front is the sharpest

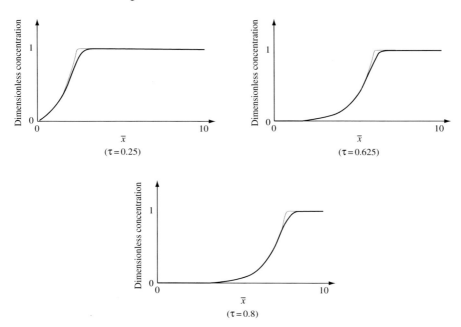

Fig. 5.3 Comparison of numerical solutions with analytical ones at different time instants (Dimensionless concentration): the *thick line* shows the numerical results, while the *thin line* shows the corresponding analytical solutions

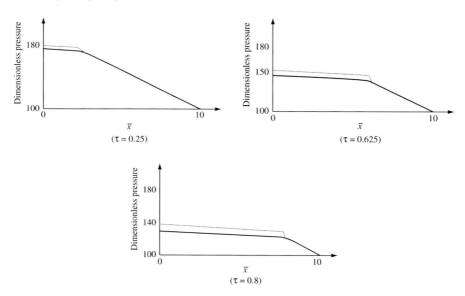

Fig. 5.4 Comparison of numerical solutions with analytical ones at different time instants (Dimensionless pressure): the *thick line* shows the numerical results, while the *thin line* shows the corresponding analytical solutions

one among the three propagation fronts, namely a porosity propagation front, a dimensionless-concentration propagation front and a dimensionless-pressure propagation front, in the computational model. Clearly, the dimensionless-pressure propagation front has the widest bandwidth, implying that it is the least sharp front in the computational model. Although there are some smoothing effects on the numerically-simulated propagation fronts as a result of numerical dispersion, the propagation speed of the numerically-simulated propagation front is in good coincidence with that of the analytically-predicted propagation front. For this benchmark problem, the overall accuracy of the numerical results is indicated by the dimensionless pressure. The maximum relative error of the numerically-simulated dimensionless pressure is 2.2%, 4.6% and 5.8% for dimensionless times of 0.25, 0.625 and 0.8 respectively. If both a small mesh size and a small time step are used, then the maximum relative error can be further reduced in the numerical simulation. This quantitatively demonstrates that the proposed numerical procedure can produce accurate numerical solutions for the planar dissolution-front propagation problem within a fluid-saturated porous medium.

5.3 Application of the Segregated Algorithm for Simulating the Morphological Evolution of Chemical Dissolution Fronts

In this section, the proposed numerical procedure is used to simulate the morphological evolution of a chemical dissolution front in a supercritical system. For this purpose, a dimensionless-pressure gradient (i.e. $\overline{p}'_{fx} = -10$) is applied on the left boundary of the computational domain so that the dimensionless speed of the dissolution front propagation is equal to 100. This means that the dissolution front propagates much faster than it does within the system considered in the previous section. Due to this change, the ratio of the equilibrium concentration to the solid molar density of the chemical species is assumed to be 0.001, while the dimensionless time-step is also assumed to be 0.001 in the computation. The Zhao number of the system is increased to 10, which is greater than the critical Zhao number (i.e. approximately 1.77) of the system. The values of other parameters are exactly the same as those used in the previous section. Since the Zhao number of the system is greater than its critical value, the coupled system considered in this section is supercritical so that a planar dissolution front evolves into a complicated morphology during its propagation within the system. In order to simulate the instability of the chemical dissolution front, a small perturbation of 1% initial porosity is randomly added to the initial porosity field in the computational domain.

Figure 5.5 shows the porosity distributions due to the morphological evolution of the chemical dissolution front in the fluid-saturated porous medium, while Fig. 5.6 shows the dimensionless concentration distributions due to the morphological evolution of the chemical dissolution front within the computational domain. It is observed that for the values of the dimensionless time greater than 0.03, the initial planar dissolution front gradually changes into an irregular one. With a further increase of the dimensionless time, the amplitude of the resulting irregular

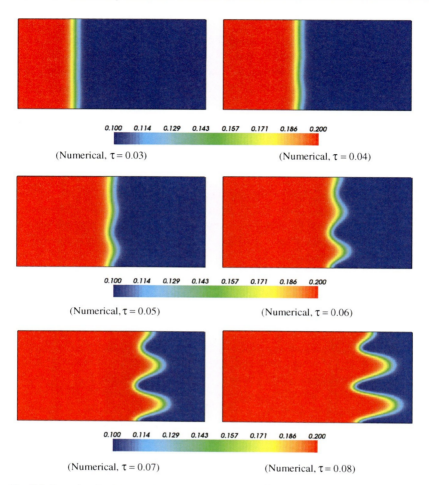

Fig. 5.5 Porosity distributions due to morphological evolution of the chemical dissolution front in the fluid-saturated porous medium

dissolution front increases significantly, indicating that the chemical dissolution front is morphologically unstable during its propagation within the computational model. Although both the porosity and the dimensionless concentration have a similar propagation front, the distribution of their maximum values along the dissolution front is clearly different. The peak value of the porosity is in good correspondence with the trough value of the dimensionless concentration due to the chemical dissolution in the system. This demonstrates that the proposed numerical procedure is capable of simulating the morphological instability of the chemical dissolution front in a fluid-saturated porous medium in the case of the coupled system being supercritical.

5.3 Application of the Segregated Algorithm

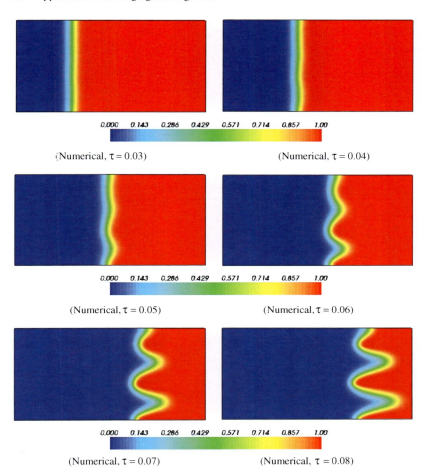

Fig. 5.6 Dimensionless concentration distributions due to morphological evolution of the chemical dissolution front in the fluid-saturated porous medium

It is interesting to investigate how the dimensionless pressure and pore-fluid flow evolve with time during propagation of the unstable dissolution front in the computational model. Figure 5.7 shows the dimensionless pressure distributions during the morphological evolution of the chemical dissolution front. It is noted that although the dimensionless pressure is continuous, there exists a clear transition for the dimensionless pressure-gradient distribution in the computational model. This phenomenon can be clearly seen at the late stages of the numerical simulation such as when the dimensionless time is equal to 0.06 and 0.07. The fluid-flow pattern evolution during the propagation of the unstable dissolution front is exhibited by the streamline evolution in the computational model. Figure 5.8 shows the

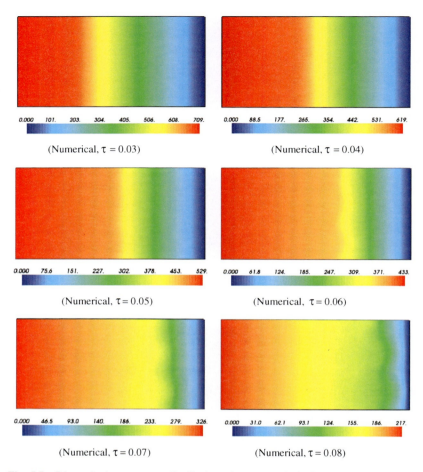

Fig. 5.7 Dimensionless pressure distributions due to morphological evolution of the chemical dissolution front in the fluid-saturated porous medium

streamline distributions during the morphological evolution of the chemical dissolution front within the coupled system between porosity, pore-fluid pressure and reactive chemical-species transport. Due to the growth of the amplitude of the irregular dissolution front, pore-fluid flow focusing takes place in the peak range of the porosity, which can be observed from the streamline density (in Fig. 5.8). It is noted that the width of the flow focusing zone is closely associated with the peak and trough values of the irregular dissolution front in the computational model. Since both the porosity generation and the pore-fluid flow focusing play an important role in ore body formation and mineralization, the proposed numerical procedure can provide a useful tool for simulating the related physical and chemical processes associated with the generation of giant ore deposits within the upper crust of the Earth.

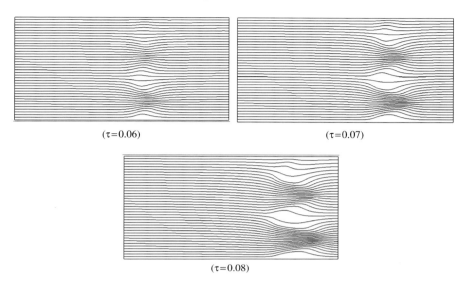

Fig. 5.8 Streamline distributions due to morphological evolution of the chemical dissolution front in the fluid-saturated porous medium

Chapter 6
A Decoupling Procedure for Simulating Fluid Mixing, Heat Transfer and Non-Equilibrium Redox Chemical Reactions in Fluid-Saturated Porous Rocks

Non-equilibrium redox chemical reactions of high orders are ubiquitous in fluid-saturated porous rocks within the crust of the Earth. They play a very important role in ore body formation and alteration closely associated with a mineralizing system. Since pore-fluid is a major carrier transporting chemical species from one part of the crust into another, the chemical process is coupled with the pore-fluid flow process in fluid-saturated porous rocks. In addition, if the rate of a chemical reaction is dependent on temperature, the chemical process is also coupled with the heat transfer process. When a pore-fluid carrying one type of chemical species meets with that carrying another type of chemical species, these two types of pore-fluids can mix together to allow the related chemical reaction to take place due to solute molecular diffusion/dispersion and advection. For these reasons, the resulting patterns of mineral dissolution, transportation, precipitation and rock alteration are a direct consequence of coupled processes between fluids mixing, heat transfer and chemical reactions in fluid-saturated porous rocks.

Due to ever-increasing demands for minerals and possible exhaustion of the existing mineral deposits in the foreseeable future, understanding the controlling mechanisms behind ore body formation and mineralization within the upper crust of the Earth becomes a very important research field. There is no doubt that an improved understanding of ore forming processes can significantly promote mineral exploration for new ore deposits within the upper crust of the Earth. Although extensive studies have been carried out to understand the possible physical and chemical processes associated with ore body formation and mineralization (Phillips 1991, Yeh and Tripathi 1991, Nield and Bejan 1992, Steefel and Lasaga 1994, Raffensperger and Garven 1995, Schafer et al. 1998a, b, Zhao et al. 1997a, 1998a, 1999b, Xu et al. 1999, Zhao et al. 2000b, 2001b, 2001d, 2002c, Schaubs and Zhao 2002, Ord et al. 2002, Gow et al. 2002, Zhao et al. 2003a), the kinetics of a redox chemical reaction and its interaction with physical processes are often overlooked in the numerical modelling of ore forming systems. In most chemical reactions associated with an ore forming system, the chemical reaction rate is finite so that an interaction between the solute molecular diffusion/dispersion, advection and chemical kinetics must be considered.

In terms of numerical modelling of coupled problems between fluids mixing, heat transfer and chemical reactions in fluid-saturated porous rocks, it is possible to divide the coupled problems into the following three categories (Zhao et al. 1998a). In the first category of coupled problem, the time scale of the advective flow is much smaller than that of the relevant chemical reaction in porous rock masses so that the rate of the chemical reaction can be essentially taken to be zero in the numerical analysis. For this reason, the first category of coupled problem is often called the non-reactive mass transport problem. In contrast, for the second category of coupled problem, the time scale of the advective flow is much larger than that of the relevant chemical reaction in pore-fluid saturated porous rocks so that the rate of the chemical reaction can be essentially taken to be infinite, at least from the mathematical point of view. This means that the equilibrium state of the chemical reaction involved is always attained in this category of coupled problem. As a result, the second category of coupled problem is called the quasi-instantaneous equilibrium problem. The intermediate case between the first and the second category belongs to the third category of coupled problem, in which the rate of the relevant chemical reaction is a positive real number of finite value. Another significant characteristic of the third category of coupled problem is that the detailed kinetics of the chemical reactions must be taken into account. It is the kinetics of a chemical reaction that describes the reaction term in a reactive species transport equation. If a redox chemical reaction is considered, both the forward reaction rate and the backward one need to be included in the reaction term of a reactive species transport equation. Although significant achievements have been made for the numerical modelling of non-reactive species and quasi-instantaneous equilibrium reaction transport problems, research on the numerical modelling of the third category of coupled problem with redox chemical reactions is rather limited. Considering this fact, we will develop a numerical procedure to solve coupled problems between fluids mixing, heat transfer and redox chemical reactions in fluid-saturated porous rocks.

Large geological faults and cracks are favorable locations for fluids carrying different chemical species to focus and mix. For this reason, ore body formation and mineralization are often associated with geological faults and cracks. When the permeability of a fault/crack is much bigger than that of the surrounding rock, the pore-fluid flow in the fault/crack is much faster than that in the surrounding rock. This implies that an interaction between the solute diffusion, advection and chemical kinetics is very strong within and around a fault/crack. Although it is well known that ore body formation and mineralization are associated with geological faults and cracks, the major factors controlling the reaction patterns within and around large faults and cracks remains unclear.

Keeping the above-mentioned considerations in mind, a numerical approach based on the finite element method is used to solve coupled problems between fluids mixing, heat transfer and redox chemical reactions in fluid-saturated porous rocks. In order to improve the efficiency of numerical modelling, the concept of the chemical reaction rate invariant is used to convert the conventional reactive transport equations with strong chemical reaction terms into the following two different kinds of equations: One is the same as the first category of mass transport equation without any reaction term; while another remains the same as the third category of reactive transport equation with a strong reaction term. Since the solution of a reactive transport equation with a chemical reaction term is computationally much more expensive than that of a mass transport equation without a chemical reaction term, any reduction in the total number of reactive transport equations can significantly save computer time in a numerical computation. Based on this idea, a decoupling numerical procedure, which can be used to remove the coupling between redox types of reactive transport equations, is proposed to solve coupled problems between fluids mixing, heat transfer and redox chemical reactions in fluid-saturated porous rocks. This allows the interaction between the solute molecular diffusion/dispersion, advection and chemical kinetics to be investigated within and around faults and cracks in the upper crust of the Earth.

6.1 Statement of Coupled Problems between Fluids Mixing, Heat Transfer and Redox Chemical Reactions

For pore-fluid saturated porous rocks, Darcy's law can be used to describe pore-fluid flow and the Oberbeck-Boussinesq approximation is employed to describe a change in pore-fluid density due to a change in the pore-fluid temperature. Fourier's law and Fick's law can be used to describe the heat transfer and mass transport phenomena respectively. If the pore-fluid is assumed to be incompressible, the governing equations of the coupled problem between fluids mixing, heat transfer and redox chemical reactions in fluid-saturated porous rocks can be expressed as follows:

$$\frac{\partial u}{\partial x} + \frac{\partial v}{\partial y} = 0, \tag{6.1}$$

$$u = \frac{K_x}{\mu}\left(-\frac{\partial P}{\partial x}\right), \tag{6.2}$$

$$v = \frac{K_y}{\mu}\left(-\frac{\partial P}{\partial y} + \rho_f g\right), \tag{6.3}$$

$$[\phi \rho_f c_{pf} + (1-\phi)\rho_s c_{ps}]\frac{\partial T}{\partial t} + \rho_f c_{pf}\left(u\frac{\partial T}{\partial x} + v\frac{\partial T}{\partial y}\right) = \lambda_{ex}\frac{\partial^2 T}{\partial x^2} + \lambda_{ey}\frac{\partial^2 T}{\partial y^2}, \tag{6.4}$$

$$\phi\frac{\partial C_i}{\partial t} + \left(u\frac{\partial C_i}{\partial x} + v\frac{\partial C_i}{\partial y}\right) = \left(D_{ex}\frac{\partial^2 C_i}{\partial x^2} + D_{ey}\frac{\partial^2 C_i}{\partial y^2}\right) + \phi R_i \quad (i = 1, 2, \dots, N), \tag{6.5}$$

$$\rho_f = \rho_{f0}[1 - \beta_T(T - T_0)], \tag{6.6}$$

$$\lambda_{ex} = \phi\lambda_{fx} + (1-\phi)\lambda_{sx}, \qquad \lambda_{ey} = \phi\lambda_{fy} + (1-\phi)\lambda_{sy}, \qquad (6.7)$$

$$D_{ex} = \phi D_{fx}, \qquad D_{ey} = \phi D_{fy}, \qquad (6.8)$$

where u and v are the horizontal and vertical velocity components of the pore-fluid in the x and y directions respectively; P is the pore-fluid pressure; T is the temperature of the porous medium; C_i is the concentration of chemical species i; N is the total number of the active chemical species considered in the pore-fluid; K_x and K_y are the permeabilities of the porous medium in the x and y directions respectively; μ is the dynamic viscosity of the pore-fluid; ρ_f and ρ_s are the densities of the pore-fluid and solid matrix; g is the acceleration due to gravity; ρ_{f0} and T_0 are the reference density and reference temperature used in the analysis; λ_{fx} and λ_{sx} are the thermal conductivities of the pore-fluid and solid matrix in the x direction; λ_{fy} and λ_{sy} are the thermal conductivities of the pore-fluid and solid matrix in the y direction; c_{pf} and c_{ps} are the specific heat of the pore-fluid and solid matrix respectively; D_{fx} and D_{fx} are the diffusivities of the chemical species in the x and y directions respectively; ϕ is the porosity of the porous medium; β_T is the thermal volume expansion coefficient of the pore-fluid; R_i is the source/sink term for the reactive transport equation of chemical species i.

It is noted that if the aqueous mineral concentrations associated with ore body formation and mineralization are relatively small, their contributions to the density of the pore-fluid are negligible so that the mass transport process can be decoupled from the pore-fluid flow and heat transfer processes. This means that the whole coupled problem between fluids mixing, heat transfer and redox chemical reactions in fluid-saturated porous rocks can be divided into two new problems. The first is a coupled problem between the pore-fluid flow and the heat transfer process, while the second is a coupled problem between the mass transport process and the redox chemical reaction process. Since the first coupled problem, which is described by Eqs. (6.1), (6.2), (6.3) and (6.4), can be solved using the existing finite element method (Lewis and Schrefler 1998, Zienkiewicz 1977), the main purpose of this study is to develop a new decoupling procedure to effectively and efficiently solve the second coupled problem, which is described by Eq. (6.5) and the related chemical reaction equations.

If the reaction term in Eq. (6.5) can be determined and is linearly dependent on the chemical species concentration, then the coupled problem defined between fluids mixing, heat transfer and redox chemical reactions in fluid-saturated porous rocks above is solvable using the numerical methods available (Zhao et al. 1998a, 2003a). This requires that the chemical reaction be of the first order. Since many chemical reactions of different orders are associated with ore body formation and mineralization in fluid-saturated porous rocks, both the second order and the high order chemical reactions are very common in nature. Without loss of generality, the second order redox chemical reaction is considered in order to develop a concept resulting in a new decoupling procedure for removing the coupling between reactive transport equations of redox chemical reactions. In principle, the new concept

6.1 Problems between Fluids Mixing, Heat Transfer and Redox Chemical Reactions

and decoupling procedure can be extended to deal with the high order redox chemical reactions. For this reason, a redox chemical reaction of the second order is considered as follows:

$$A + B \underset{k_b}{\overset{k_f}{\Leftrightarrow}} AB, \tag{6.9}$$

where A and B are two chemical reactants; AB is the chemical product due to this redox chemical reaction; k_f and k_b are the forward and backward reaction rates of this redox chemical reaction. It needs to be pointed out that Eq. (6.9) represents a class of redox chemical reactions such as $H^+ + OH^- \underset{k_b}{\overset{k_f}{\Leftrightarrow}} H_2O$, $Na^+ + Cl^- \underset{k_b}{\overset{k_f}{\Leftrightarrow}} NaCl$, $Ca^{2+} + CO_3^{2-} \underset{k_b}{\overset{k_f}{\Leftrightarrow}} CaCO_3$ and so forth in geochemical systems. It is clear that in the first chemical reaction example, chemical reactants A and B are H^+ and OH^-, while chemical product AB is H_2O. In the second chemical reaction example, chemical reactants A and B are Na^+ and Cl^-, while chemical product AB is $NaCl$. Similarly, in the third chemical reaction example, chemical reactants A and B are Ca^{2+} and CO_3^{2-}, while chemical product AB is $CaCO_3$.

From the chemical reaction point of view, the general chemical reaction source/sink terms due to the redox chemical reaction expressed by Eq. (6.9) can be written as follows:

$$R_A = -k_f r^{n_f - 1} C_A C_B + k_b r^{n_b - 1} C_{AB}, \tag{6.10}$$

$$R_B = -k_f r^{n_f - 1} C_A C_B + k_b r^{n_b - 1} C_{AB}, \tag{6.11}$$

$$R_{AB} = k_f r^{n_f - 1} C_A C_B - k_b r^{n_b - 1} C_{AB}, \tag{6.12}$$

where C_A, C_B and C_{AB} are the concentrations of chemical species A, B and AB; R_A, R_B and R_{AB} are the chemical reaction source/sink terms associated with chemical species A, B and AB; n_f and n_b are the orders of the forward and backward reactions respectively; r is a quantity of unity value to balance the unit of the reaction source/sink terms due to different orders of chemical reactions so that it has a reciprocal unit of the chemical species concentration. For the redox reaction expressed by Eq. (6.9), the forward reaction is of the second order, while the backward reaction is of the first order. Since a redox system allows chemical reactions to be proceeded toward both the product and the reactant directions, the orders of the forward reaction (i.e. the chemical reaction proceeds toward the product direction) and backward reaction (i.e. the chemical reaction proceeds toward the reactant direction) can be determined from the related chemical kinetics.

It is noted that the accumulation or diffusion of chemical species in the rock matrix may lead to some change in porosity, which in turn affects permeability and fluid flow in the rock matrix (Zhao et al. 2001d, Xu et al. 2004). This influence can be straightforwardly considered using variable permeability within the computational model. The permeability change induced by a chemical reaction can be determined from the porosity variation induced by the chemical reaction. For example, the Carman-Kozeny law can be used to establish a relationship between the chemically induced porosity change and the chemically induced permeability change in the rock matrix.

6.2 A Decoupling Procedure for Removing the Coupling between Reactive Transport Equations of Redox Chemical Reactions

Substituting Eqs. (6.10), (6.11) and (6.12) into Eq. (6.5) yields the following equations:

$$\phi \frac{\partial C_A}{\partial t} + \left(u \frac{\partial C_A}{\partial x} + v \frac{\partial C_A}{\partial y} \right) = \left(D_{ex} \frac{\partial^2 C_A}{\partial x^2} + D_{ey} \frac{\partial^2 C_A}{\partial y^2} \right) + \phi R_A, \quad (6.13)$$

$$\phi \frac{\partial C_B}{\partial t} + \left(u \frac{\partial C_B}{\partial x} + v \frac{\partial C_B}{\partial y} \right) = \left(D_{ex} \frac{\partial^2 C_B}{\partial x^2} + D_{ey} \frac{\partial^2 C_B}{\partial y^2} \right) + \phi R_B, \quad (6.14)$$

$$\phi \frac{\partial C_{AB}}{\partial t} + \left(u \frac{\partial C_{AB}}{\partial x} + v \frac{\partial C_{AB}}{\partial y} \right) = \left(D_{ex} \frac{\partial^2 C_{AB}}{\partial x^2} + D_{ey} \frac{\partial^2 C_{AB}}{\partial y^2} \right) + \phi R_{AB}. \quad (6.15)$$

Since the total number of linearly-independent reaction rates is identical to the total number of chemical reactions involved, there is only one linearly-independent reaction rate for this redox chemical reaction. From this point of view, the total number of reactive transport equations with source/sink terms due to chemical reactions can be reduced into one, for this particular redox chemical system. This is the basic idea behind the proposed decoupling procedure for removing the coupling between reactive transport equations of redox chemical reactions.

Through some algebraic manipulations, Eqs. (6.13), (6.14) and (6.15) can be rewritten as follows:

$$\phi \frac{\partial (C_A + C_{AB})}{\partial t} + \left[u \frac{\partial (C_A + C_{AB})}{\partial x} + v \frac{\partial (C_A + C_{AB})}{\partial y} \right] \\ - \left[D_{ex} \frac{\partial^2 (C_A + C_{AB})}{\partial x^2} + D_{ey} \frac{\partial^2 (C_A + C_{AB})}{\partial y^2} \right] = 0, \quad (6.16)$$

$$\phi \frac{\partial (C_B + C_{AB})}{\partial t} + \left[u \frac{\partial (C_B + C_{AB})}{\partial x} + v \frac{\partial (C_B + C_{AB})}{\partial y} \right] \\ - \left[D_{ex} \frac{\partial^2 (C_B + C_{AB})}{\partial x^2} + D_{ey} \frac{\partial^2 (C_B + C_{AB})}{\partial y^2} \right] = 0, \quad (6.17)$$

6.2 A Decoupling Procedure for Removing the Coupling

$$\phi \frac{\partial C_{AB}}{\partial t} + \left[u \frac{\partial C_{AB}}{\partial x} + v \frac{\partial C_{AB}}{\partial y} \right] - \left[D_{ex} \frac{\partial^2 C_{AB}}{\partial x^2} + D_{ey} \frac{\partial^2 C_{AB}}{\partial y^2} \right]$$
$$= \phi k_b \left(\frac{k_f}{k_b} r^{n_f - 1} C_A C_B - r^{n_b - 1} C_{AB} \right).$$
(6.18)

Equations (6.16) and (6.17) are two conventional mass transport equations without any source/sink terms due to the redox chemical reaction so that they can be solved by the well-developed numerical methods available. Since the two new variables, namely $C_I = C_A + C_{AB}$ and $C_{II} = C_A + C_{AB}$, are independent of chemical reaction rates, they can be referred to as chemical reaction rate invariants, which are the analogues of the stress and strain invariants in the field of solid mechanics.

If the redox chemical reaction is an equilibrium one, then both the forward and the backward reaction rates are theoretically infinite so that the chemical reaction becomes predominant in the reactive transport process. In this case, Eq. (6.18) can be written as

$$K_e r^{n_f - 1} C_A C_B - r^{n_b - 1} C_{AB} = 0,$$
(6.19)

where $K_e = k_f / k_b$ is the chemical equilibrium constant.

Inserting the two chemical reaction rate invariants, $C_I = C_A + C_{AB}$ and $C_{II} = C_B + C_{AB}$, into Eq. (6.19) yields the following equation:

$$K_e r (C_I - C_{AB})(C_{II} - C_{AB}) - C_{AB} = 0.$$
(6.20)

It is noted that $n_f = 2$ and $n_b = 1$ are substituted into Eq. (6.19) so as to obtain Eq. (6.20). Clearly, Eq. (6.20) has the following mathematical solution for the chemical product of the redox chemical reaction:

$$C_{AB} = \frac{K_e r (C_I + C_{II}) + 1}{2 K_e r} - \sqrt{\frac{[K_e r (C_I + C_{II}) + 1]^2 - 4 K_e^2 r^2 C_I C_{II}}{4 K_e^2 r^2}}.$$
(6.21)

This indicates that for an equilibrium chemical reaction, we only need to solve the mass transport equations of chemical reaction rate invariants (i.e. $C_I = C_A + C_{AB}$ and $C_{II} = C_B + C_{AB}$ in this particular example) using the conventional numerical methods. Once the distributions of the chemical reaction rate invariants are obtained in a computational domain, the chemical product distribution due to the chemical reaction can be calculated analytically. As a result, the distributions of the chemical reactants can be calculated using the distributions of both the chemical product and the chemical reaction rate invariants.

However, for non-equilibrium chemical reactions, the chemical reaction rates are finite so that we need to solve at least one reactive transport equation with the source/sink term for each of the chemical reactions in the geochemical system. This means that for the general form of the redox chemical reaction considered in this study, we need to solve Eq. (6.18) numerically if the reaction rates of this redox chemical reaction are finite. For this reason, Equation (6.18) can be rewritten into the following form:

$$\phi \frac{\partial C_{AB}}{\partial t} + \left[u \frac{\partial C_{AB}}{\partial x} + v \frac{\partial C_{AB}}{\partial y} \right] - \left[D_{ex} \frac{\partial^2 C_{AB}}{\partial x^2} + D_{ey} \frac{\partial^2 C_{AB}}{\partial y^2} \right] \\ - \phi k_b \left\{ K_e r C_{AB}^2 - [K_e r(C_I + C_{II}) + 1] C_{AB} + K_e r C_I C_{II} \right\} = 0. \quad (6.22)$$

Since Eq. (6.22) is strongly nonlinear with the nonlinear term, $\phi k_b K_e r C_{AB}^2$, the Newton-Raphson algorithm is suitable for solving this equation.

Using the proposed decoupling procedure, the coupled problem between fluids mixing, heat transfer and redox chemical reactions in fluid-saturated porous rocks can be solved in the following five main steps: (1) For a given time step, the coupled problem described by Eqs. (6.1), (6.2), (6.3), (6.4) and (6.5) with the related boundary and initial conditions are solved using the conventional finite element method; (2) After the pore-fluid velocities are obtained from the first step, mass transport equations of the chemical reaction rate invariants (i.e. Eqs. (6.16) and (6.17)) with the related boundary and initial conditions are then solved using the existing finite element method; (3) The chemical reaction source/sink terms involved in Eq. (6.22) is determined from the related redox chemical reaction so that Eq. (6.22) can be solved using the Newton-Raphson algorithm; (4) According to the definitions of the chemical reaction rate invariants, $C_I = C_A + C_{AB}$ and $C_{II} = C_B + C_{AB}$, the chemical reactant concentrations (i.e. C_A and C_B) can be determined from simple algebraic operations; (5) Steps (1–4) are repeated until the desired time step is reached. These solution steps have been programmed into our research code.

6.3 Verification of the Decoupling Procedure

The main and ultimate purpose of a numerical simulation is to provide numerical solutions for practical problems in a real world. Since numerical methods are the basic foundation of a numerical simulation, only can an approximate solution be obtained from a computational model, which is the discretized description of a continuum mathematical model. Due to inevitable round-off errors in computation and discretized errors in temporal and spatial variables, it is necessary to verify, at least from the qualitative point of view, the proposed numerical procedure so that meaningful numerical results can be obtained from a discretized computational model. For this reason, a testing coupled problem, for some aspects of which the analytical solutions are available, is considered in this section.

Figure 6.1 shows the geometry of the coupled problem between pore-fluids mixing, heat transfer and redox chemical reactions around a vertical geological fault within the crust of the Earth. For this problem, the pore-fluid pressure is assumed to be lithostatic, implying that there is an upward throughflow at the bottom of the computational model. The height and width of the computational model are 10 km

6.3 Verification of the Decoupling Procedure

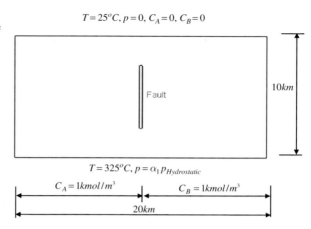

Fig. 6.1 Geometry and boundary conditions of the fluid focusing and mixing problem

and 20 km respectively. The length of the fault is 5 km, with an aspect ratio of 20. The porosities of the fault and its surrounding rock are 0.35 and 0.1. The surrounding rock is assumed to have a permeability of 10^{-16} m^2 in both the x and y directions, while the permeability of the fault is calculated using the Carman-Kozeny formula, which gives rise to a permeability of about 43 times that of the surrounding rock. The top temperature is 25°C and the geothermal gradient is 30°C/km, meaning that the temperature at the bottom is fixed at 325°C. The two chemical reactants with a concentration of 1 kmol/m^3 are injected at the left half and right half bottom boundary respectively, while the concentrations of both the two reactants and the product are assumed to be zero at the top boundary of the computational model. The dispersion/diffusivity of the chemical species is 3×10^{-10} m^2/s. For the pore-fluid, dynamic viscosity is 10^{-3} N × s/m^2; reference density is 1000 kg/m^3; volumetric thermal expansion coefficient is 2.1×10^{-4} (1/°C); specific heat is 4184 J/(kg × °C); thermal conductivity coefficient is 0.59 W/(m × °C) in both the x and y directions. For the porous matrix, the thermal conductivity coefficient is 2.9 W/(m × °C) in both the x and y directions; the specific heat is 878 J/(kg × °C); reference rock density is 2600 kg/m^3. In order to simulate the fluids focusing and chemical reactions within the fault in an appropriate manner, a mesh of small element sizes is used to simulate the fault zone, while a mesh gradation scheme is used to simulate the surrounding rock by gradually changing the mesh size from the outline of the fault within the computational model. As a result, the whole computational domain is simulated by 306,417 three-node triangle elements.

Figure 6.2 shows the streamline distribution of the system with the vertical fault in the computational model. Since the pore-fluids carrying two different chemical reactants are uniformly and vertically injected into the computational model at the left and right parts of the bottom, the pore-fluid flow converges into the vertical fault at the inlet (i.e. the lower end) of the fault, but diverges out of the vertical fault at the outlet (i.e. the upper end) of the fault. This phenomenon can be clearly observed from Fig. 6.3, where the velocity distributions are displayed at both the inlet and

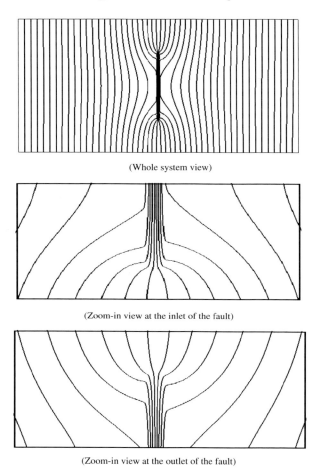

(Whole system view)

(Zoom-in view at the inlet of the fault)

(Zoom-in view at the outlet of the fault)

Fig. 6.2 Streamline distributions due to the fluid focusing within a vertical fault in the crust

outlet of the vertical fault. For an elongate elliptic fault of large aspect ratios in an infinite medium, the existing analytical solution indicates that the streamlines of the pore-fluid flow are parallel each other inside of the fault (Zhao et al. 1999d). As expected, the computed streamlines concentrate vertically within the vertical fault, indicating that the numerical result obtained from the computational model has good agreement with the existing analytical result. This demonstrates that the proposed numerical procedure can produce useful numerical solutions for fluid focusing and mixing, at least from the qualitative point of view. In order quantitatively to validate the numerical solutions, the analytical solution for the flow-focusing factor of an elongate elliptic fault of large aspect ratios can be employed. Since the elongate fault within the computational model is basically of a rectangular shape, the analytical solution for its flow-focusing factor can be evaluated using the following modified formula:

6.3 Verification of the Decoupling Procedure

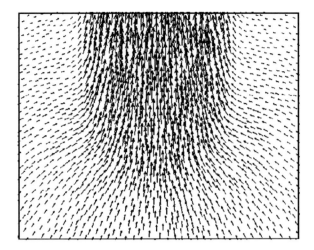

(Zoom-in view at the inlet of the fault)

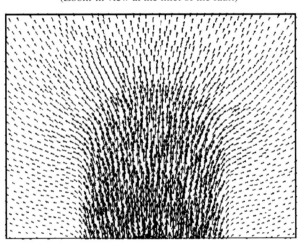

(Zoom-in view at the outlet of the fault)

Fig. 6.3 Velocity distributions due to the fluids focusing within a vertical fault in the crust

$$\lambda = \left(\frac{4}{\pi}\right) \frac{\alpha(\beta + 1)}{\beta + \alpha}, \tag{6.23}$$

where λ is the pore-fluid flow focusing factor of the rectangular fault of a large aspect ratio; β is the aspect ratio of the rectangular fault; α is the permeability ratio of the fault to its surrounding rock.

Since the vertical velocity of the injected fluids due to the lithostatic pore-fluid pressure is equal to 1.6×10^{-9} m/s and the numerical solution for the maximum vertical velocity within the fault is equal to 3.02×10^{-8} m/s, the corresponding flow-focusing factor of the fault is equal to the ratio of the maximum velocity within the fault to that of the injected fluids at the bottom of the computational model. This results in a flow-focusing factor of 18.88 for the rectangular fault within the computational model. The analytical value of the flow-focusing factor for the rectangular fault can be calculated from Eq. (6.23). Substituting the related parameters into Eq. (6.23) yields the analytical flow-focusing factor of 18.25. Since the relative error of the flow-focusing factor from the numerical simulation is within 3.5%, it quantitatively demonstrates that the proposed numerical procedure used in the computational model can produce accurate numerical solutions for fluid focusing and mixing within the fault. These processes are very important for accurately simulating the chemical species transport and reaction within the fault.

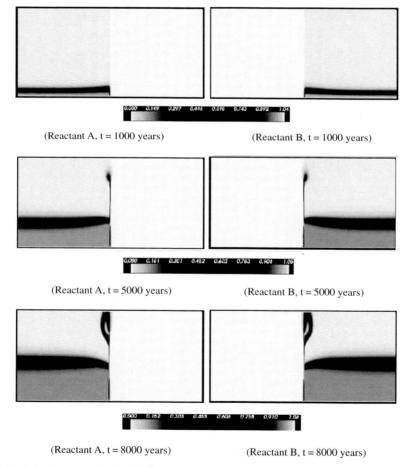

(Reactant A, t = 1000 years)   (Reactant B, t = 1000 years)

(Reactant A, t = 5000 years)   (Reactant B, t = 5000 years)

(Reactant A, t = 8000 years)   (Reactant B, t = 8000 years)

Fig. 6.4 Concentration distributions of the chemical reactants at different time instants (Equilibrium reaction)

6.3 Verification of the Decoupling Procedure

In order to verify the proposed decoupling procedure for solving reactive transport equations with strong nonlinear reaction source/sink terms, the redox chemical reaction due to an equilibrium reaction is considered and solved using the proposed numerical procedure. The equilibrium constant is assumed to be 10 and the time step used in the simulation is 100 years. Figure 6.4 shows the concentration distributions of the two chemical reactants at three different time instants, while Fig. 6.5 shows the comparison of the numerical solutions (which are obtained from the proposed decoupling procedure) with the analytical solutions (which are derived mathematically and expressed by Eq. (6.21)) for the chemical product. It can be observed that with the increase of time, both chemical reactants are transported into the computational domain from the left half and right half of the bottom. Due to the fluid flow focusing, both chemical reactants are transported much faster in the fault zone than in the surrounding rock. As expected, these chemical reactants

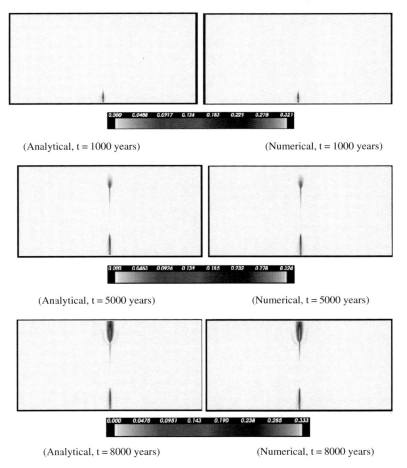

Fig. 6.5 Comparison of numerical solutions with analytical ones for the chemical product (Equilibrium reaction)

are divergent around the exit region of the fault. The comparison of the numerical solutions with the analytical ones for the chemical product clearly demonstrates that the proposed decoupling procedure can produce accurate numerical solutions for simulating the equilibrium chemical reaction, in which the chemical reaction rate approaches infinite. It is interesting to note that there is a strong interaction between solute advection, diffusion and chemical reaction rate in the considered equilibrium chemical system. Although two reactants are well transported into the fault zone, the mixing of the two fluids carrying them is controlled by the solute diffusion. Since the chemical reaction rate is infinite for the equilibrium reaction, the corresponding chemical equilibrium length is identical to zero due to the solute diffusion. This implies that the chemical reaction rate is too fast to allow both the reactants to diffuse across the common boundary between them so that fluid mixing cannot effectively take place within the fault zone. This is the reason why both chemical reactants are abundant but no chemical product is produced within the fault zone in the computational model. However, around the exit region of the fault zone, the flow of the fluids is slowed and divergent so that the fluids carrying two different reactants can be mixed. Consequently, a high concentration of the chemical product is produced around the exit region of the fault zone.

6.4 Applications of the Proposed Decoupling Procedure to Predict Mineral Precipitation Patterns in a Focusing and Mixing System Involving Two Reactive Fluids

Mixing of two or more fluids is commonly suggested as a mechanism for precipitating minerals from solution in porous rocks. Examples include uranium deposits (Wilde and Wall 1987), MVT deposits (Appold and Garven 2000, Garven et al. 1999), Irish Pb-Zn deposits (Hitzmann, 1995, Everett et al. 1999, Murphy et al. 2008), vein gold deposits (Matthai et al. 1995, Cox et al. 1995) and Carlin gold deposits (Cline and Hofstra 2000). The mixing process is attractive because it enables two fluids of contrasting Eh-pH conditions to mix and hence generate chemical conditions conducive to mineral precipitation. Clearly rock-fluid(s) interactions must be involved as well as the fluid mixing process but in depth exploration of such multi-fluid-rock reaction processes awaits robust and computationally fast ways of handling realistic kinetic-reaction-transport phenomena.

There are three end member geometries that promote the mixing of two miscible fluids (Fig. 6.1). The first we refer to as parallel-flow geometries of the first kind (Figs. 6.6a, b) where two contrasting fluids are brought alongside each other by convection or focusing in a highly permeable fault or sedimentary lens (Phillips 1991). The second involves the production of a new fluid within an existing fluid flow system by thermal or chemical reactions with a mineral assemblage within the flow stream (Fig. 6.6c). An example is the production of CO_2 or of hydrocarbons from carbonates or carbonaceous material within a fluid stream that is hot, acid and

6.4 Applications of the Proposed Decoupling Procedure

oxidised (Matthai et al. 1995, Cox et al. 1995). This is, in fact, a special example of parallel-flow geometry except that now the new fluid source is embedded within the primary flow stream. We refer to this process as parallel flow geometry of the second kind. The third end member involves injection of one fluid perpendicular to the flow field of a second fluid as shown in Figs. 6.6d, e. This mixing geometry produces a mixing plume that expands outwards mainly in the downstream direction of the

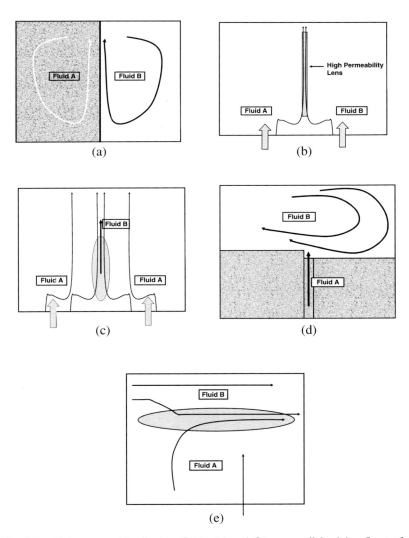

Fig. 6.6 Mixing geometries for two fluids: (**a**) and (**b**) are parallel mixing flows of the type discussed in this paper; (**c**) is a parallel mixing flow but the second fluid is generated within the flow field of the first fluid; (**d**) and (**e**) are perpendicular mixing flows where one fluid is injected normal to the flow field of the other fluid

second flow. This situation has been considered by Phillips (1991) and analytical solutions for the composition of the resulting mixing plume presented.

6.4.1 Key Factors Controlling Mineral Precipitation Patterns in a Focusing and Mixing System Involving Two Reactive Fluids

We consider the first of these parallel flow geometries in some detail. The classical approach (see Phillips 1991) is to assume that chemical equilibrium has been established within the mixing plume so that the equilibrium concentration of a particular chemical species is expressed as a function of the environmental parameters, namely temperature, fluid pressure and the concentration of other chemical species. The fundamental point we explore here is that the overall pattern of mineralization in these mixing systems results from intimate interactions between solute advection, solute diffusion and/or dispersion and chemical kinetics. Thus, chemical equilibrium in some cases may never be attained in these mixing systems.

In this section we concentrate solely on parallel flows of the first kind and show that the patterns of chemical precipitation for such flows result from intimate relations between fluid flow (which is a first order control on solute advection), chemical reaction rates, and solute diffusion and dispersion. We show, for instance, that even for a situation where two fluids are brought together by fluid focusing within a permeable fault zone, chemical reaction between the two fluids may never occur even if chemical equilibrium prevails so long as the advection of solute is large with respect to solute diffusion/dispersion. In order to focus on the principles involved for parallel flows of the first kind we idealise the situation by considering a vertical permeable fault within a less permeable rock mass. The whole model is fully saturated with fluids of different chemical compositions at different points on the system. We consider only systems where the fluids are miscible whilst noting for future investigation that multi-phase systems, that is, fluid systems in which two or more fluids are present that are immiscible, comprise yet another hydraulic-chemical system where intimate mixing of contrasting chemistries is possible at the pore scale. We need to emphasise here that although we are considering only vertical permeable faults the discussion is just as relevant to permeable sedimentary lenses or to any other geometry that brings two chemically contrasted fluids together. The vertical geometry here allows simple boundary conditions to be imposed for the hydraulic potential driving fluid flow. The discussion is equally relevant for any other orientation of the focusing "lens".

Ore formation in the Earth's crust is commonly considered to be a consequence of the reaction of large fluxes of one or more fluids with rock. In many situations, fluid flow is focused within fault zones and the important issue becomes the spatial control exerted by the fault on the resultant mineralization. It is commonly assumed that either fluid mixing or fluid/rock reactions or a combination of both are important processes associated with permeable fault zones, due to the large fluid fluxes typically inferred for such zones. However, recent extensive studies (Phillips 1991,

6.4 Applications of the Proposed Decoupling Procedure

Zhao et al. 1998a, 2007a) have demonstrated that ore formation is not only dependent on fluid flow, but also upon solute diffusion/dispersion and chemical kinetics.

Mineralisation is commonly associated with the spatial distribution of chemical reactions in permeable rocks (Steefel and Lasaga 1990, Phillips 1991, Raffensperger and Garven 1995, Zhao et al. 1998a). These chemical reaction patterns are strongly influenced by fluid flow, heat transfer and the transport of reactive chemical species. Since fluid flow is an important agent for transporting aqueous chemical species from one location into another (Garven and Freeze 1984, Peacock 1989, Ord and Oliver 1997, Connolly 1997, Zhao et al. 1999d, Oliver 2001, Braun et al. 2003), fluid flow patterns can also influence the positions where different chemical species may meet and mix and hence where chemical reactions may take place (Yardley and Bottrell 1992, Yardley and Lloyd 1995). In addition, heat transfer processes influence the thermal structure of the Earth's crust and hence the spatial distribution of chemical reaction patterns through the temperature-dependence of the equilibrium constant and of the reaction rate (Steefel and Lasaga 1994, Xu et al. 1999, Zhao et al. 2000b).

For aqueous species, whether chemical equilibrium is or is not attained controls the possibility of precipitation of mineral assemblages in permeable rocks. If an aqueous species, produced by a chemical reaction, reaches equilibrium, then that species is saturated within the fluid and hence its concentration reaches a maximum value for a given mineralizing environment. In most cases, the solubility of an aqueous species is directly proportional to the ambient temperature. It follows that in cooling hydrothermal systems the saturated aqueous species becomes oversaturated and is therefore precipitated in the porous rock. Another process causing mineral precipitation is transport within the pore-fluid of the saturated aqueous species from a high temperature region to a low temperature region. The aqueous species can also become oversaturated and is precipitated in the porous rock. For a given ore formation environment, flow rates can control the total process of solute advection. In such a case, a redox controlled chemical reaction can only reach equilibrium through optimal coupling of solute advection, solute diffusion/dispersion and chemical kinetics. With parallel flows taken as an example, if the solute diffusion/dispersion in the direction perpendicular to the interface between the flows is much slower than the chemical reaction rate, the chemical product cannot attain equilibrium, even though the two fluids are perfectly miscible. On the other hand, if the flow rate is much faster than the chemical reaction rate, the chemical product can only attain equilibrium at a distance large with respect to a chemical equilibrium length-scale, which we introduce below, in the flow direction. This implies that for a permeable fault zone with a given chemical reaction, there exists an optimal flow rate resulting in chemical equilibrium being attained between two fluids that mix and focus within the fault zone. However, for parallel flows, such as those resulting from vertical super-hydrostatic pressure gradients, chemical equilibrium may or may not be attained when two fluids of different origins are brought together through focusing within a permeable vertical fault. In this case, strong interactions occur between solute advection, diffusion/dispersion and chemical kinetics. It is this interaction that controls the equilibrium distribution of the resulting chemical product. A conceptual model is presented here to investigate this interaction and the corresponding influence on chemical reaction patterns.

6.4.2 Theoretical Analysis of Mineral Precipitation Patterns in a Focusing and Mixing System Involving Two Reactive Fluids

Although it is difficult, if not impossible, to obtain analytical solutions for the coupled problem expressed by Eqs. (6.1), (6.2), (6.3), (6.4) and (6.5) in general cases, it is possible to gain theoretical understanding through analytical solutions for the coupled problem in some limiting cases (Zhao et al. 2007a). This understanding can be achieved by converting the reactive transport equation (i.e. Eq. (6.5)) into a dimensionless one so that the controlling processes associated with the reactive chemical species transport can be identified. This means that it is possible to investigate the relationship between the controlling processes associated with reactive chemical species transport so that the overall structure of their solutions can be understood. Since there are three major controlling processes, namely solute advection, solute diffusion/dispersion and chemical kinetics, that may play dominant roles in determining chemical reaction patterns, we need to consider the relationships between the time scales for these three controlling processes. For this purpose, we first consider a dimensionless parameter known as the Damköhler number, Da, (Steefel and Lasaga 1990, Ormond and Ortoleva 2000) to express the relative time scales of solute advection and reaction kinetics:

$$Da = \frac{\phi k_R l}{V} = \textit{Time Scale for Solute Advection}/ \quad (6.24)$$
$$\textit{Time Scale for Chemical Reaction},$$

where V is the characteristic Darcy velocity of the system; l is the characteristic length of the controlling process in the system; k_R is the controlling chemical reaction rate with units of [s^{-1}]; ϕ is the porosity of the porous medium. When the time scale for solute advection is equal to the time scale for chemical kinetics, the Damköhler number is equal to one. In this case, the chemical equilibrium length scale of the system can be expressed as follows:

$$l_{advection}^{chemical} = \frac{V}{\phi k_R}, \quad (6.25)$$

where $l_{advection}^{chemical}$ is the chemical equilibrium length due to solute advection for a given chemical reaction. Below we refer to $l_{advection}^{chemical}$ as *the advection chemical equilibrium length*. It is clear that if a chemical reaction rate is given, there exists an optimal flow rate such that the chemical reaction can reach equilibrium beyond the advection chemical equilibrium length determined from Eq. (6.25). Thus, for a given $l_{advection}^{chemical}$, the corresponding optimal flow rate, $V_{optimal}$, for which the chemical reaction can reach equilibrium beyond the given $l_{advection}^{chemical}$, is as follows:

6.4 Applications of the Proposed Decoupling Procedure

$$V_{optimal} = \phi k_R l_{advection}^{chemical}. \tag{6.26}$$

It should be noted that both $l_{advection}^{chemical}$ and the optimal flow-rate have clear physical meanings: The physical meaning of $l_{advection}^{chemical}$ is that for a given fluid flow rate, a chemical reaction with a given reaction rate can reach equilibrium once this distance is traversed by the fluid in the flow direction, within the time scale of chemical equilibrium. Since $l_{advection}^{chemical}$ is directly proportional to the fluid flow rate, the greater the fluid flow rate, the larger $l_{advection}^{chemical}$. This means that fast flows require relatively long distances in the flow direction, beyond which a chemical reaction with a given reaction rate can reach equilibrium. In contrast, the physical meaning of the optimal flow rate is that for a given $l_{advection}^{chemical}$, a chemical reaction with a given reaction rate can reach equilibrium if the fluid flow rate is within the time scale of chemical equilibrium. Since the optimal flow rate is also directly proportional to $l_{advection}^{chemical}$, the larger $l_{advection}^{chemical}$ in the flow direction, the greater the optimal fluid flow rate. This implies that a large $l_{advection}^{chemical}$ requires a relatively fast optimal flow rate, so that chemical reaction for a given reaction rate can reach equilibrium beyond this large $l_{advection}^{chemical}$ in the flow direction.

If the flow paths of two fluids are parallel to each other in a fluid mixing system, solute diffusion/dispersion normal to the flow direction plays a fundamental role in promoting chemical reactions between different reactive chemical species. In this case, a second dimensionless parameter, Z, needs to be defined to express the relative time scale between the solute diffusion/dispersion process and the chemical reaction process. Notice that Z is independent of the fluid velocity and so still has meaning for zero fluid flow.

$$Z = \frac{k_R l^2}{D} = \textit{(Time Scale for Solute Dispersion/Diffusion)}/ \tag{6.27}$$

Time Scale for Chemical Reaction,

where D is the solute diffusion/dispersion coefficient; l is the characteristic length of the controlling process in the system; k_R is the reaction rate. Since this dimensionless number expresses the ratio of the solute diffusion/dispersion time scale to the chemical kinetic time scale, it is unity when the two time scales are equal. In this situation, the chemical equilibrium length, $l_{diffusion}^{chemical}$ of the system can be expressed as follows:

$$l_{diffusion}^{chemical} = \sqrt{\frac{D}{k_R}}, \tag{6.28}$$

where $l_{diffusion}^{chemical}$ is the *chemical equilibrium length due to solute diffusion/dispersion* for a given chemical reaction. For a given solute diffusion/dispersion coefficient, there exists an optimal reaction rate such that the chemical reaction can reach equilibrium within $l_{diffusion}^{chemical}$ determined from Eq. (6.28). Thus, for a given $l_{diffusion}^{chemical}$, the corresponding optimal chemical reaction rate, $k_R^{optimal}$, for which the chemical reaction can reach equilibrium within the given $l_{diffusion}^{chemical}$, is as follows:

$$k_R^{optimal} = \frac{D}{\left(l_{diffusion}^{chemical}\right)^2}. \tag{6.29}$$

Both $l_{diffusion}^{chemical}$ and the optimal chemical reaction-rate also have clear physical meanings: The physical meaning of $l_{diffusion}^{chemical}$ is that for parallel flow with a given solute diffusion/dispersion coefficient, a chemical reaction with a given reaction rate can only reach equilibrium within the distance diffused by the solute in the direction perpendicular to the fluid flow within the time scale of chemical equilibrium. Since $l_{diffusion}^{chemical}$ is directly proportional to the square root of the solute diffusion/dispersion coefficient, the greater the solute diffusion/dispersion coefficient, the larger $l_{diffusion}^{chemical}$ in the direction perpendicular to the fluid flow. This implies that a large solute diffusion/dispersion coefficient can result in a chemical reaction, with a given reaction rate, reaching equilibrium over a large distance normal to the direction of fluid flow. By contrast, the physical meaning of the optimal chemical reaction rate is that for a given $l_{diffusion}^{chemical}$, a chemical reaction with a given reaction rate can reach equilibrium if the chemical reaction proceeds at this chemical reaction rate within the time scale of chemical equilibrium. Since the optimal chemical reaction rate is inversely proportional to the square of $l_{diffusion}^{chemical}$, the larger $l_{diffusion}^{chemical}$ in the direction perpendicular to the fluid flow, the smaller the optimal chemical reaction rate. This implies that a large chemical equilibrium length due to solute diffusion/dispersion in the direction perpendicular to the parallel fluid flow requires a relatively slow optimal chemical reaction rate, so that for a given solute diffusion/dispersion coefficient, the chemical reaction can reach equilibrium within this long chemical equilibrium length in the direction perpendicular to the parallel fluid flow.

The combined use of these two numbers, Da and Z, can express the interaction between solute advection, diffusion/dispersion and chemical kinetics. Note that for a given fault zone involving pore-fluid flow focusing and mixing, it is possible to define three different types of mineral precipitation patterns in Z-Da number space. For this purpose, the thickness of the fault zone is chosen as the characteristic length of the Z Number, while the length of the fault zone is chosen as the characteristic length of the Da Number. The relationships between Da, Z and mineral precipitation types are shown in Fig. 6.7.

6.4.3 Chemical Reaction Patterns due to Mixing and Focusing of Two Reactive Fluids in Permeable Fault Zones

The theoretical understanding of the interaction between solute advection, solute diffusion/dispersion and chemical kinetics presented in the previous section is, in principle, useful for investigating chemical reaction patterns resulting from chemical equilibrium associated with fluid flow in all kinds of porous rocks. Chemical reaction patterns arising from flow focusing and mixing of two reactive fluids within permeable vertical fault zones are the subject of this section.

6.4 Applications of the Proposed Decoupling Procedure

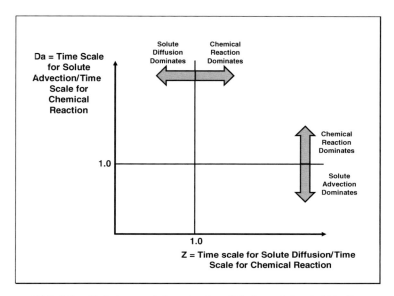

(A) Relationship between solution advection, diffusion and chemical kinetics

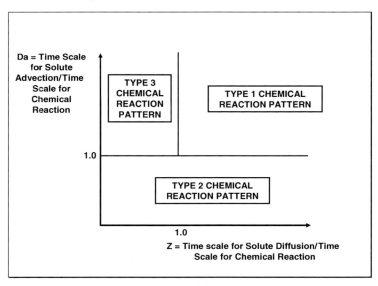

(B) Three types of mineral precipitation patterns

Fig. 6.7 Three fundamental types of mineral precipitation patterns in the Z-Da Number space

The geometry and problem description of two fluids mixing and focusing within a permeable vertical fault zone are shown in Fig. 6.8. In this figure, L_{fault} and W_{fault} are the length and width of the fault zone; L_{mp} represents the starting position of mineral precipitation within the fault zone; W_{mp} represents the thickness of mineral

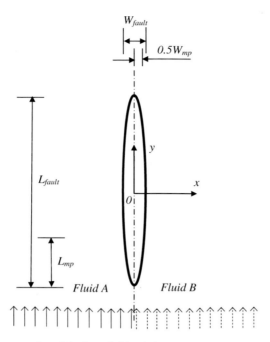

Fig. 6.8 The conceptual model of two fluids mixing and focusing in a fault

precipitation within the fault zone. Two fluids carrying two different solutes, namely Fluid *A* and Fluid *B*, are injected from the left and right sides of the fault axis and are focused into the fault zone according to the principles discussed by Phillips (1991) and Zhao et al. (1999d). Since mineral precipitation is dependent on both chemical equilibrium and the saturation concentration of a mineral, it is assumed that the starting position of mineral precipitation is coincident with chemical equilibrium being attained in the following theoretical analysis. This means that for a given chemical reaction rate, the starting position of mineral precipitation is controlled by solute advection and is measured from the lower tip of the fault. On the other hand, since Fluid *A* flows parallel to Fluid *B*, the mixing of these two fluids is due solely to solute diffusion/dispersion in the direction normal to the fault zone. Hence, the thickness of mineral precipitation is controlled solely by solute diffusion/dispersion. The starting position of mineral precipitation within the fault zone can be expressed as follows:

$$L_{mp} = l_{advection}^{chemical} = \frac{V}{\phi k_R}. \tag{6.30}$$

Similarly, the thickness of mineral precipitation within the fault zone can be expressed as

$$W_{mp} = 2l_{diffusion}^{chemical} = 2\sqrt{\frac{D}{k_R}}. \tag{6.31}$$

6.4 Applications of the Proposed Decoupling Procedure

Here the factor 2 arises because of the symmetry of geometry; $l_{diffusion}^{chemical}$ is measured from the centre of the fault.

As indicated earlier in this chapter, there are three possible cases in which chemical equilibrium can be attained in the conceptual model shown in Fig. 6.8. The three types of interest are: Type 1: chemical equilibrium is attained just at the lower tip of the fault; Type 2: chemical equilibrium is attained at the upper tip of the fault; and Type 3: chemical equilibrium is attained somewhere between the lower tip and upper tip of the fault. As mentioned above, analytical solutions to Eqs. (6.1), (6.2), (6.3), (6.4) and (6.5) are difficult but we can gain some insight into the types of chemical reaction patterns that are possible by considering some limiting cases below. In the forthcoming section, we numerically examine more realistic and applicable situations.

6.4.3.1 Type 1: Chemical Equilibrium is Attained at the Lower Tip of the Fault

In this limiting case, the fluid velocity is zero in the whole system. Substituting a zero velocity into Eq. (6.30) yields a zero value of the starting position of mineral precipitation within the fault zone. In this situation, it is possible for chemical equilibrium to be attained throughout the whole length of the fault, depending on the actual residence time of the two chemical reactants within the system. To enable chemical equilibrium to be reached within the whole fault length, L_{fault}, the residence time for which the two chemical reactants should exist in the system is equal to the time interval, $t_{diffuse}$, for the solute to diffuse from the lower tip to the upper tip of the fault:

$$t_{diffuse} = \frac{L_{fault}^2}{D}. \tag{6.32}$$

Consider a specific example: Eq. (6.32) indicates that for a solute diffusion/dispersion coefficient of 10^{-10} m^2/s and a vertical fault 1 km long, the two chemical reactants need to be present for 10^{16} s (i.e. about 3×10^8 years) at the bottom of the conceptual model in order for chemical equilibrium to be reached within the whole length of the fault. Since the required duration for the existence of these two chemical reactants is quadratically proportional to the length of a fault, it is increased to about 25×10^{16} s (i.e. about 7.5×10^9 years) to enable the chemical reaction to reach equilibrium within the whole length of a 5 km long fault. In this example, where the fluid flow is negligible, the time required to reach chemical equilibrium within the whole length of a fault 5 km long is greater than the age of the Earth.

As indicated in Eq. (6.31), the thickness of mineral precipitation within the fault zone is dependent on both the chemical reaction rate and the solute diffusion/dispersion coefficient. For a given solute diffusion/dispersion coefficient, the optimal chemical reaction rate $k_R^{optimal}$, which will enable chemical reactions to attain equilibrium across the whole width of the fault zone, is as follows:

$$k_R^{optimal} = \frac{4D}{W_{fault}^2}. \tag{6.33}$$

6.4.3.2 Type 2: Chemical Equilibrium is Attained at the Upper Tip of the Fault

This is another limiting case, in which the fluid rate (i.e. velocity) has reached a critical value in the fault zone. The fundamental characteristic of this limiting case is that if the fluid rate is equal to or greater than this critical value, the starting point of chemical equilibrium is at or beyond the upper tip of the fault, indicating that the chemical equilibrium of the chemical product and therefore mineral precipitation cannot be attained within the fault zone. For a given chemical reaction rate, this critical flow rate within the fault zone, $V_{critical}$, can be expressed as follows:

$$V_{critical} = \phi k_R L_{fault}. \tag{6.34}$$

Equation (6.34) is useful for estimating the critical flow rate for a vertical fault. For example, if the porosity of a vertical fault is 0.3 and the chemical reaction rate is 10^{-11}(1/s), then the corresponding critical flow rate is 3×10^{-9} m/s for a vertical fault 1 km long. If both the fault porosity and the chemical reaction rate remain unchanged, the corresponding critical flow rate of the fault is increased to about 15×10^{-9} m/s for a vertical fault 5 km long. Suppose the flow-focusing factor (Phillips, 1991) of the vertical fault is of the order of 15, the corresponding critical flow rate within the surrounding rock for a vertical fault is 10^{-9} m/s. This critical flow rate can be easily exceeded when the fluid within the surrounding rock of the fault zone is under a lithostatic pressure gradient. If the permeability of the surrounding rock is 10^{-16} m^2 and the dynamic viscosity of the fluid is 10^{-3} N s/m^2, then the flow rate induced by the lithostatic pressure gradient within the surrounding rock is 1.7×10^{-9} m/s. This implies that the flow rate induced by a lithostatic pressure gradient within the surrounding rock is too high to enable minerals to be precipitated within the vertical fault zone.

6.4.3.3 Type 3: Chemical Equilibrium is Attained Somewhere between the Lower Tip and Upper Tip of the Fault

In this case, the equilibrium of the reaction product is attained within the fault zone. The starting position of the chemical equilibrium is measured from the lower tip of the fault and can be expressed by the following equation:

$$L_{mp} = \alpha_{mp} L_{fault}, \tag{6.35}$$

where L_{mp} is the starting position of chemical equilibrium within the fault zone in the flow direction; α_{mp} is a coefficient to express the relative relationship between the length of the fault and the starting position of chemical equilibrium within the fault zone; that is $\alpha_{mp} = L_{mp}/L_{fault}$. It is clear that in this particular case, the value

6.4 Applications of the Proposed Decoupling Procedure

of α_{mp} is greater than zero but less than one. A value of α_{mp} equal to zero represents the first limiting case (Type 1), while a value of α_{mp} equal to one represents the second limiting case (Type 2), as discussed above.

Similarly, the width of chemical equilibrium within the fault can be expressed as follows:

$$W_{mp} = \beta_{mp} W_{fault}, \qquad (6.36)$$

where W_{mp} is the width of chemical equilibrium within the fault zone; W_{fault} is the width of the fault; β_{mp} is a coefficient expressing the ratio between the width of the fault and the width of chemical equilibrium within the fault zone; that is, $\beta_{mp} = W_{mp}/W_{fault}$. In this case, the value of β_{mp} is greater than zero but less than one. A value of β_{mp} equal to zero represents the first limiting case (Type 2), whilst a value of β_{mp} equal to one represents another limiting case (i.e. the limiting case of Type 3), in which case the width of chemical equilibrium within the fault zone is just equal to that of the fault.

If both the solute diffusion/dispersion coefficient and the width of chemical equilibrium within the fault zone are known, the corresponding optimal chemical reaction rate in this case can be estimated from the following formula:

$$k_R^{optimal} = \frac{4D}{W_{mp}^2} = \frac{4D}{\beta_{mp}^2 W_{fault}^2}. \qquad (6.37)$$

Thus, the corresponding optimal flow rate in this particular case is:

$$V_{optimal} = \phi k_R^{optimal} L_{mp} = \frac{4\phi D \alpha_{mp} L_{fault}}{\beta_{mp}^2 W_{fault}^2}. \qquad (6.38)$$

Equation (6.38) indicates that for a particular pattern of mineral precipitation within the fault zone, there exists an optimal flow rate such that chemical equilibrium is attained within the region associated with this particular mineral precipitation pattern.

6.4.4 Numerical Illustration of Three Types of Chemical Reaction Patterns Associated with Permeable Fault Zones

The theoretical analysis carried out in the previous section predicts that there exist three fundamental types of chemical reaction patterns associated with permeable vertical fault zones due to two fluids mixing and focusing. In order to illustrate these different types of chemical reaction patterns, the finite element method is used to solve the coupled problem numerically involving fluid mixing, heat transfer and chemical reactions expressed by Eqs. (6.1), (6.2), (6.3), (6.4) and (6.5). In theory, it is possible to predict exactly the three fundamental types of chemical reaction patterns. However, in numerical practice, it is difficult to control the flow rate of a specific value due to numerical round off and cutoff errors. For this reason, the numerical simulated mineral precipitation patterns here are approximate representations of the three fundamental types of mineral precipitation patterns.

The computational model (shown in Fig. 6.1) and related parameters for the verification example in the previous section (i.e. Sect. 6.3) are used to illustrate three different types of mineral precipitation in a focusing and mixing system of two reactive fluids. Two different fluid pressure gradients are considered to control the flow rate (i.e. vertical fluid velocity) within the surrounding rock during the numerical simulation. To simulate the fast flow rate associated with type 2 (as stated in the previous section), the fluid pressure gradient is assumed to be a lithostatic pressure gradient, which results in a vertical flow rate of 1.7×10^{-9} m/s within the surrounding rock. This gradient is used because it is considered to be near the highest reasonable fluid pressure gradient likely to be encountered in nature. On the other hand, in order to simulate the slow flow rate associated with types 1 and 3 (see the previous section), the excess fluid pressure gradient is assumed to be one percent of a lithostatic pressure gradient minus a hydrostatic one, which results in a vertical flow rate of 1.7×10^{-11} m/s within the surrounding rock. It is known that for fluid pressure gradient dominated flow, the flow pattern around a permeable fault zone is dependent only on the contrast in permeability between the fault and the surrounding rocks, the geometry of the fault zone and the inflow direction relative to the axis of the fault (Phillips 1991, Zhao et al. 1999d) and is independent of the pressure gradient along the fault. This means that although the fluid pressure gradient is different in the three simulations, the streamline pattern within and around the fault zone must be identical for all three cases, as demonstrated by the related numerical result shown in Fig. 6.2. In all cases, fluid flow converges into the fault zone at the lower end and diverges out of the fault zone at the upper end (see also Phillips 1991, Zhao et al. 1999d).

6.4.4.1 The First Type of Chemical Reaction Pattern

The first type of chemical reaction pattern results from an approximate representation of Type 1 in Sect. 6.4.3; the background flow velocity within the surrounding rock is not exactly equal to zero in the numerical simulation but instead is one percent of the lithostatic pressure gradient minus the hydrostatic gradient. Here two controlling chemical reaction rates, namely 10^{-7} (1/s) and 10^{-11} (1/s) are considered to investigate the effect on chemical reaction patterns. Due to flow focusing, the maximum vertical flow velocity within the fault zone is about 3.16×10^{-10} m/s.

Figure 6.9 shows the concentration distributions of the two chemical reactants and the corresponding chemical product at two time instants of $t = 500,000$ and $t = 800,000$ years. The distribution of the concentration of the chemical product comprises a lenticular shape within the fault zone. This coincides with what is expected from the previous theoretical analysis. For a fault zone of width 250 m, the optimal chemical reaction rate calculated from Eq. (6.33) is 4.8×10^{-15} (1/s) so that chemical equilibrium can be attained within the whole width of the fault zone. Since the optimal chemical reaction rate is much smaller than the two chemical

6.4 Applications of the Proposed Decoupling Procedure

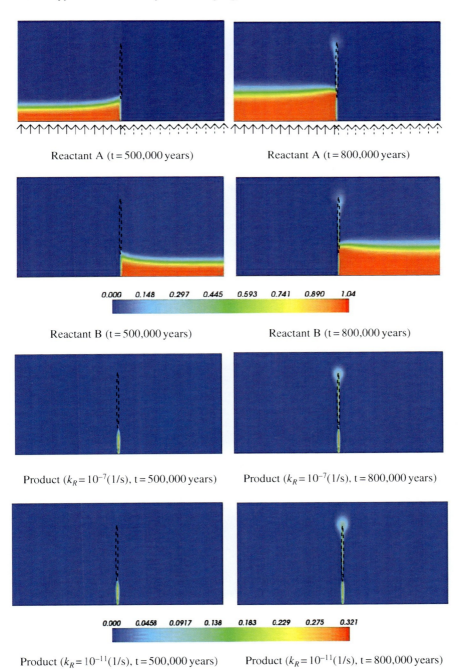

Fig. 6.9 Concentration distributions of the chemical reactants and product at two different time instants (Type 1)

reaction rates used in the simulation, the thickness of the chemical equilibrium within the fault zone is much smaller than the width of the fault zone itself. The theoretical estimate of the thickness of chemical equilibrium within the fault zone is 10.95 m for a controlling chemical reaction rate of 10^{-11} (1/s), and 0.11 m for a controlling chemical reaction rate of 10^{-7} (1/s). Hence, the maximum concentration distribution of the chemical product comprises a very thin membrane; that is, mineral precipitation comprises a thin lenticular shape within the fault zone and starting at the lower tip of the fault. This is the fundamental characteristic of the first type of chemical reaction pattern.

6.4.4.2 The Second Type of Chemical Reaction Pattern

The second type of chemical reaction pattern is defined as an approximate representation of Type 2 in the previous section. In this example, the background fluid pressure gradient within the surrounding rock is set equal to a lithostatic pressure gradient. Two controlling chemical reaction rates, namely $k_R = \infty$ and $k_R = 10^{-11}(1/s)$, are considered to investigate the effects of different chemical kinetics on chemical reaction patterns. From a chemical kinetics point of view, $k_R = \infty$ represents an equilibrium chemical reaction. Due to flow focusing, the maximum vertical flow velocity within the fault zone is about 3.02×10^{-8} m/s.

Figure 6.10 shows the concentration distributions for the two chemical reactants and the corresponding chemical product at two time instants of $t = 5000$ and $t = 8000$ years. Both chemical reactants are transported into the computational domain from the left half and right half of the base of the model. Due to fluid flow focusing, both chemical reactants are transported much faster within the fault zone than in the surrounding rock. There is a strong interaction between solute advection, diffusion/dispersion and chemical reaction rate (i.e. $k_R = \infty$). Although both reactants are transported into the fault zone, the mixing of the two fluids carrying them is controlled by solute diffusion and dispersion. Since the chemical reaction rate is infinite, the corresponding chemical equilibrium length due to solute diffusion/dispersion is identical to zero normal to the fault zone. This implies that the controlling chemical reaction rate is too fast to allow both the reactants to diffuse across their common boundary so that fluid mixing cannot effectively take place within the fault zone. However, around the exit region of the fault zone, the fluid flow decreases and, more importantly, diverges so that a high concentration of the chemical product is produced around the exit region of the fault zone.

In the case of a non-equilibrium chemical reaction characterized by a slow chemical reaction rate (e.g. $k_R = 10^{-11}(1/s)$), the theoretical chemical equilibrium length calculated from Eq. (6.28) is 5.48 m, while the theoretical chemical equilibrium length calculated from Equation (6.25) is 8630 m in the direction of the fault axis. Thus the chemical equilibrium length due to solute advection is greater than the length of the fault itself, so that the chemical product distribution within the fault zone is controlled by solute advection. Because the chemical equilibrium length due to the solute advection is greater than the total length of the fault zone plus its exit region (i.e. 5000 m plus 2500 m), chemical equilibrium cannot be reached within the

6.4 Applications of the Proposed Decoupling Procedure

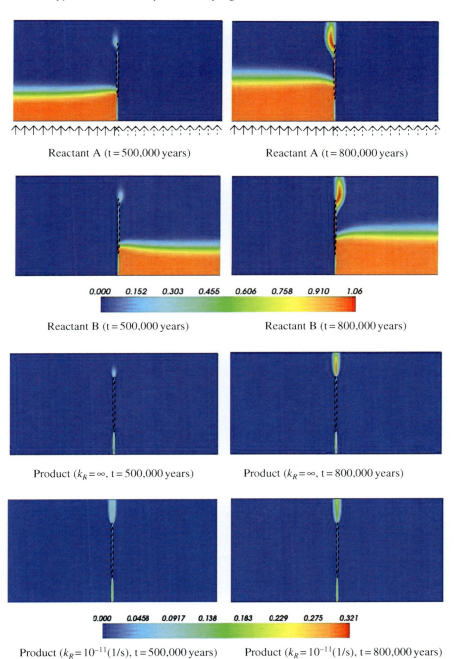

Fig. 6.10 Concentration distributions of the chemical reactants and product at two different time instants (Type 2)

height of the model, indicating that two fluids cannot mix to produce an extensive mixing region due to the fast advection of the fluids within the fault.

Thus, for the two end-members of chemical reaction rates, namely a very fast equilibrium reaction and a slow reaction with a controlling chemical reaction rate of 10^{-11} (1/s), chemical equilibrium cannot be attained within the fault, implying that mineral precipitation cannot take place within a permeable vertical fault zone for the chemical reaction considered here. This is the fundamental characteristic of the second type of chemical reaction pattern considered in this investigation.

6.4.4.3 The Third Type of Chemical Reaction Pattern

The third type of chemical reaction pattern is a representation of Type 3 in the previous section. The fundamental characteristic of this type of chemical reaction pattern is that chemical equilibrium, which results in a considerable equilibrium thickness in the direction normal to the fault zone, can be achieved within the fault zone. In order to enable the starting position of chemical equilibrium to be close to the lower tip of the vertical fault, the background fluid pressure gradient within the surrounding rock is one percent of the lithostatic pressure gradient minus the hydrostatic gradient. Due to flow focusing, the maximum vertical flow velocity within the fault zone is about 4.68×10^{-10} m/s in the numerical simulation. Since the optimal reaction rate is directly proportional to the solute diffusion/dispersion coefficient, it is desirable to select the value of a solute diffusion/dispersion coefficient as large as possible, so that the total CPU time in the simulation can be significantly reduced. For this reason, the solute diffusion/dispersion coefficient is assumed to be 3×10^{-8} m^2/s in the numerical simulation. If the thickness of chemical equilibrium in the lateral direction of the fault zone is 80 m, the optimal chemical reaction rate is about 1.9×10^{-11} (1/s), as can be calculated from Eq. (6.37).

Figure 6.11 shows concentration distributions of the two chemical reactants and the corresponding chemical product at two time instants of $t = 500,000$ and $t = 800,000$ years. The maximum concentration distribution of the chemical product generates considerable thickness within the fault zone. This agrees well with what is expected from the previous theoretical analysis given in Sect. 6.4.3. Due to the low fluid velocity, a chemical equilibrium zone is also generated in the flow convergent region just in front of the fault zone. It is noted that this chemical equilibrium zone is almost separated from the chemical equilibrium zone within the fault. This phenomenon results from the distribution of fluid velocity vectors arising from fluid flow focusing just outside and within the fault zone.

The geological implication of the third type of chemical reaction pattern is that if a mineral precipitation pattern of a certain thickness and length within a permeable vertical fault zone is observed, then it is possible to estimate both the optimal flow rate and the optimal reaction rate during the formation of this precipitation pattern. On the other hand, if the flow rate and reaction rate are known, then it is possible to estimate the width of the potential mineral precipitation pattern within a permeable vertical fault zone.

6.4 Applications of the Proposed Decoupling Procedure

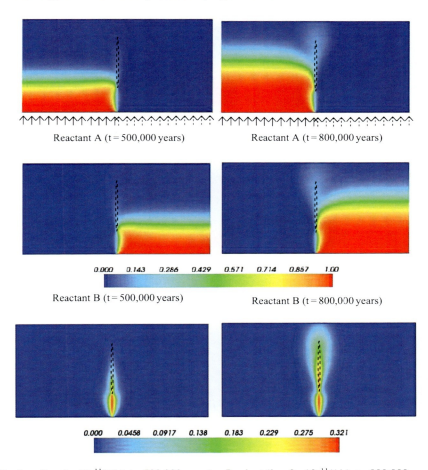

Fig. 6.11 Concentration distributions of the chemical reactants and product at two different time instants (Type 3)

Chapter 7
An Equivalent Source Algorithm for Simulating Thermal and Chemical Effects of Intruded Magma Solidification Problems

Consideration of the effects of magma ascending and solidification is important to the further understanding of ore body formation and mineralization in the crust of the Earth. Although various possible fundamental mechanisms of magma ascending in the crust are proposed (Johnson and Pollard, 1973, Marsh 1982, Lister and Kerr 1991, Rubin 1995, Weinberg 1996, Bons et al. 2001), the development of numerical algorithms for simulating the proposed magma ascending mechanisms is still under-developed. For example, continuum-mechanics-based numerical methods have encountered serious difficulties in simulating the random generation and propagation of hydro-fractured cracks, the magma flow within these cracks, the solidification of the ascending magma due to heat losses to the surrounding rocks, and so forth. In order to overcome these difficulties, particle-based numerical simulations have been developed rapidly in recent years (Zhao et al. 2006f, 2007b, c, d, 2008g). However, due to the different time and length scales involved in ore body formation and mineralization problems, it is also very difficult, even if not impossible, to use the present particle-based numerical methods to simulate all the important processes associated with ore body formation and mineralization problems in the crust of the Earth. As a long-term development strategy, we need to develop multiple time and length scale modelling techniques and algorithms so that particle simulation methods, combined with newly-developed techniques and algorithms, can be used to solve such large scale geological problems. As an expedient strategy, although it is impossible to use the continuum-mechanics-based numerical methods to simulate directly the magma ascent processes, we can develop some useful algorithms, in combination with continuum-mechanics-based numerical methods, to simulate the effects of the magma ascent processes. Thus, the main motivation of carrying out this study is to develop a useful algorithm to consider the dynamic consequences involved in magma ascent processes using continuum-mechanics-based numerical methods.

In terms of the magma intrusion mechanism, a large amount of theoretical work has been carried out previously, even though it is based on simple conceptual models (Johnson and Pollard, 1973, Marsh 1982, Lister and Kerr 1991, Rubin 1995, Weinberg 1996, Bons et al. 2001). Although the previous theoretical work needs to be quantitatively refined, it can be used to estimate the total volume of the intruded magma. If an analytical estimation of the volume of the intruded magma is not available for most complicated geological situations, the particle numerical method may be useful to simulate the magma intrusion process alone so that the volume of the intruded magma can also be estimated. Given the total volume of the intruded magma, we can use continuum-mechanics-based numerical methods, such as the finite element method and finite difference method, to simulate its thermal effects by considering the heat release during the solidification of the intruded magma. This means that we need to develop an equivalent algorithm to transform the original magma intrusion problem into a heat transfer problem with the internal heat generation of the intruded magma. Clearly, the key issue associated with the developed equivalent algorithm is to determine the heat release rate of the intruded magma during its solidification. Once the time history of the heat release rate of the intruded magma during solidification is obtained, the finite element method can be used to simulate the thermal effects of the intruded magma in the crust of the Earth. Using the proposed equivalent algorithm, the moving boundary problem (Crank 1984, Alexiades and Solomon 1993) associated with the original problem during magma solidification can be avoided. As a direct result, the efficiency of the finite element method can be much improved. This may be considered as one of the major advantages of the proposed equivalent algorithm in dealing with the thermal effects of magma intrusion problems in the crust of the Earth.

Geological problems may involve different time and length scales in the descriptions of their different physical and chemical processes. With magma intrusion into the Earth's crust taken as an example, the time scale of the magma intrusion process, which includes both the creation of the magma chamber and ascent process for the intruded magma, is much smaller than that of the magma solidification process, which includes both the release of volatile fluids from the magma and chemical reaction processes within the crust due to the release of the volatile fluids. On the other hand, the volume of the intruded magma is usually much smaller than that of the Earth's crust of interest. This means that the whole magma intrusion problem is, in essence, a problem of multiple time and length scales. Nevertheless, due to the significant time and length scale differences between the magma intrusion and solidification processes, it is possible to simulate these two different processes using different analytical models. This will allow the detailed mechanisms associated with each of the two processes to be modelled using totally different methodologies. For example, if the magma intrusion process itself is of particular interest, then particle-based numerical methods can be used to simulate the initiation and propagation of random cracks during the ascending of the intruded magma. However, if the thermal and chemical effects/consequences of the intruded magma are of particular interest, then continuum-mechanics-based numerical methods can be used to simulate heat transfer and mass (i.e. chemical species) transport within the Earth's crust. To the best knowledge of the authors, these methods have not been used to simulate the release of volatile fluids from the magma and the chemical reaction processes within the crust due to the release of the volatile fluids. Since both the release of volatile fluids from the magma and the transport of the released volatile fluids may have significant effects on ore body formation and mineralization in the upper crust of the Earth, a numerical method needs to be developed to simulate the chemical consequences of the magma solidification in porous rocks.

A large amount of geochemical research has indicated that intruded magma in the Earth's crust has different chemical compositions and, therefore, different types of rocks can be formed during the solidification and crystallization of the intruded magma. Both temperature and pressure conditions during solidification of the magma can also affect the resulting rock types dramatically. Rhyolite and basalt, instead of granite and gabbro, are formed if the solidification conditions of the felsic and mafic magmas are changed from intrusive into extrusive. It is well known that the solidus of the intruded magma is mainly dependent on the magma composition, the contents of water and other volatile fluids. For a particular kind of intruded magma, the abovementioned information is obtainable from geochemical and isotopic analyses. Therefore, given a particular kind of intruded magma, it is possible to determine the contents of water and other volatile fluids, which should be released when the intruded magma becomes solidified. This implies that the volatile fluids released from the intruded magma can be quantitatively simulated in the numerical analysis, which is another important issue to be addressed in this study.

Due to the complex nature of the magma intrusion problem within the crust of the Earth, it is useful to conduct numerical simulation of this kind of problem progressively from a simplistic stage into more complicated stages. This research methodology is rational because of the multiple time and length scales of the problem itself. Thus, we plan to solve the magma intrusion problem using the following three level models. For the first level model, we primarily consider the effects of the *post-solidification* magma on pore-fluid flow, heat transfer and ore forming patterns within the upper crust of the Earth. For the second level model, we consider the thermal and chemical effects of the *post-intrusion* but *pre-solidification* magma on pore-fluid flow, heat transfer and ore forming patterns within the upper crust of the Earth. In this model, we must develop some useful and efficient computer algorithms to simulate the magma solidification problem. For the third level model, we will consider the *intrusion process itself* at a much smaller scale using particle mechanics-based computer algorithms. Once efficient numerical algorithms for dealing with above three level models are developed, it is possible to integrate them to simulate the whole process of the magma intrusion problem. So far, we have completed some work for the first level model (Zhao et al. 2003e). In this study, we will develop some useful and efficient computer algorithms to simulate the magma solidification problem associated with the second level model.

7.1 An Equivalent Source Algorithm for Simulating Thermal and Chemical Effects of Intruded Magma Solidification Problems

Owing to the importance for many scientific issues and technical applications, theoretical and numerical analyses of heat transfer with phase change have been carried out for more than half a century. There is an extensive literature which reports and reviews the development of this subject (Carslaw and Jaeger 1959, Crank 1984, Alexiades and Solomon 1993, Zhao and Heinrich 2002). The problem associated with numerical modelling of the thermal effect of solidification of intruded magma is that the characteristic dimension of the whole geological system under consideration is on the scale of tens and hundreds of kilometers, but the characteristic dimension of the intruded magma, such as a sill or dike (Johnson and Pollard 1973, Lister and Kerr 1991, Rubin 1995, Weinberg 1996) is on the scale of meters and tens of meters. As a result, the detailed solidification process of the intruded magma might not be important, but the thermal effect caused by the heat release during the solidification of the intruded magma is important, at least from the ore body formation and mineralization point of view. Since heat release during solidification of the intruded magma can be represented by a physically equivalent heat source, it is possible to transform the original heat transfer problem with phase change due to solidification of the intruded magma into a physically equivalent heat transfer problem without phase change but with the equivalent heat source. The above physical understanding means that from the numerical modelling point of view, we can remove the moving boundary problem associated with the detailed solidification process of the intruded magma, so that we can use a fixed finite element mesh to consider the thermal effect of the intruded magma by solving the physically equivalent heat transfer problem with an equivalent heat source. This is the basic idea behind the proposed equivalent algorithm for simulating the thermal effect of the intruded magma solidification in this study. The similar, even though not identical, approaches, such as the immersed boundary method (Beyer and LeVeque 1992), the level-set method (Osher and Sethian 1998, Sethian 1999), the segment projection method (Tornberg and Engquist 2003a) and so forth, have been used to tackle other different problems with moving interfaces or fronts (Smooke et al. 1999, Mazouchi and Homsy 2000). In terms of solving partial differential equations with delta function source terms numerically, Walden (1999), and Tornberg and Engquist (2003b) have investigated the convergence and accuracy of the numerical method. Recently, the fixed grid approach has been successfully used to solve the phase change problem associated with moisture transport in high-temperature concrete materials (Schrefler 2004, Schrefler et al. 2002, Gawin et al. 2003). In this chapter, it will be extended to the solution of the phase change problem associated with intruded magma solidification in geological systems. The proposed algorithm is only valid when the characteristic length scale of the system is much larger than that of the intruded magma, which is true for most geological systems (Johnson and Pollard 1973, Lister and Kerr 1991, Rubin 1995, Weinberg 1996).

In what follows, we will present the proposed equivalent algorithm for simulating the chemical effect of solidification of the intruded magma in geological systems. Since the volatile fluids released during solidification of the intruded magma may have many different chemical components/species, it is ideal to describe the governing equation of each chemical component separately. However, if the diffusivity of each chemical component is assumed to be identical, then the concept of the total concentration, which is the summation of the concentration of all the chemical components, can be used to describe the governing equation of the released volatile

7.1 An Equivalent Source Algorithm for Simulating Thermal and Chemical Effects

fluids. Thus, we can use one governing equation to describe variations of the total concentration of the released volatile fluids in the system. In this way, computer efforts can be reduced significantly in dealing with the intruded magma solidification problems. Because the volumetric amount of the released volatile fluids is relatively small to the intruded magma, their effects on the heat transfer process can be neglected in the analysis of the intruded magma solidification problem. This allows us to assume that thermal equilibrium between the released volatile fluids and the intruded magma/rock has been achieved. Thus, the governing equations of the original heat transfer and mass transport problem considering the phase change during the intruded magma solidification (Carslaw and Jaeger 1959, Crank 1984, Alexiades and Solomon 1993, Zhao and Heinrich 2002) can be described, for a two-dimensional problem, as

$$(\rho_R c_{pR}) \frac{\partial T_R}{\partial t} = \lambda_R \left(\frac{\partial^2 T_R}{\partial x^2} + \frac{\partial^2 T_R}{\partial y^2} \right) \quad (x, y) \in V_R, \quad (7.1)$$

$$(\rho_M c_{pM}) \frac{\partial T_M}{\partial t} = \lambda_M \left(\frac{\partial^2 T_M}{\partial x^2} + \frac{\partial^2 T_M}{\partial y^2} \right) \quad (x, y) \in V_M, \quad (7.2)$$

$$\frac{\partial C_T}{\partial t} = D \left(\frac{\partial^2 C_T}{\partial x^2} + \frac{\partial^2 C_T}{\partial y^2} \right) + \delta(x - x_I, y - y_I) Q(x, y, t) \quad (x, y) \in V_R + V_M, \quad (7.3)$$

where T_R and T_M are the temperature of the rock and intruded magma; ρ_R, c_{pR} and λ_R are the density, specific heat and thermal conductivity of the rock; ρ_M, c_{pM} and λ_M are the density, specific heat and thermal conductivity of the intruded magma; C_T is the total concentration of the released volatile fluids; D is the diffusivity of the released volatile fluids; Q is the mass source of the released volatile fluids during the intruded magma solidification; x_I and y_I are the x and y coordinate components of the interface position; δ is the delta function of values of unity and zero; V_R and V_M are the spaces occupied by the rock and intruded magma.

It is noted that, although the released volatile fluids during the intruded magma solidification are comprised of H_2O, CO_2, H_2S, HCl, HF, SO_2 and other substances (Burnham 1979, Barns 1997), H_2O is the most abundant magmatic volatile and CO_2 is the second most abundant magmatic volatile in the intruded magma. For this reason, the solubility of H_2O in silicate melts has been investigated for many years. These extensive studies (Burnham 1979, Barns 1997) have demonstrated that if the constraints of the solution model for the $NaAlSi_3O_8 - H_2O$ system are imposed, H_2O solubilities in the igneous-rock melts are essentially identical to those in $NaAlSi_3O_8$ melts. Therefore, the solubility of H_2O in the $NaAlSi_3O_8$ melt can be used to approximately determine the mass source of the released volatile fluids during the intruded magma solidification.

Since the original magma solidification problem belongs to a moving interface problem, the temperature and heat flux continuity conditions on the interface between the rock and intruded magma are as follows:

$$T_R = T_M \qquad (x_I, y_I) \in \Gamma_{RM}, \qquad (7.4)$$

$$\lambda_R \left(n_x \frac{\partial T_R}{\partial x} + n_y \frac{\partial T_R}{\partial y} \right) - \lambda_M \left(n_x \frac{\partial T_M}{\partial x} + n_y \frac{\partial T_M}{\partial y} \right)$$
$$= \rho_M [L + c_{pM}(T_{IM} - T_m)] \left(n_x \frac{\partial x_I}{\partial t} + n_y \frac{\partial y_I}{\partial t} \right) \qquad (x_I, y_I) \in \Gamma_{RM}, \qquad (7.5)$$

where x_I and y_I are the x and y coordinate components of the interface position; n_x and n_y are the x and y components of the unit normal to the interface between the rock and intruded magma; T_{IM} is the temperature of the intruded magma; T_m is the solidification temperature of the intruded magma; L is the latent heat of fusion of the intruded magma; Γ_{RM} is the interface between the rock and intruded magma.

Except for the boundary conditions on the interface between the rock and intruded magma, the boundary conditions on the other boundaries of the rock domain and intruded magma domain can be either of the Dirichlet, Neumann or mixed type (Carslaw and Jaeger 1959, Crank 1984, Alexiades and Solomon 1993, Zhao and Heinrich 2002). Since the boundary conditions on the other boundaries of the rock domain and intruded magma domain are trivial, it is not necessary to repeat them here.

In order to develop an efficient and effective numerical algorithm for dealing with solidification of intruded magma in sills and dikes, it is necessary to introduce some related concepts. As shown in Fig. 7.1, supposing the initial position of the interface between the magma and rock is at position 1, the interface moves to position 2 due to magma solidification during a time period, Δt_M. As a result, the magma solidification thickness is expressed by ΔL_M. If the length of the interface front is ΔL_R, then the magma solidification area is the product of ΔL_M and ΔL_R during the time period Δt_M. Since heat and volatile fluids are only released during magma solidification, we may consider the released heat and volatile fluids either as surface heat and mass sources in a two-dimensional problem or as volumetric heat and mass sources in a three-dimensional one. This means that it is physically possible to use

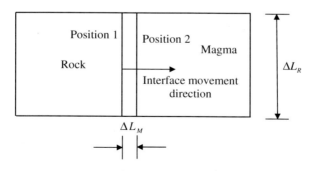

Fig. 7.1 Basic concepts related to the magma solidification problem

7.1 An Equivalent Source Algorithm for Simulating Thermal and Chemical Effects

a fixed finite element mesh to consider the released heat and volatile fluids during the magma solidification process. The key issue associated with the finite element analysis using a fixed mesh is that, for a given magma solidification thickness ΔL_M, which can be easily simulated by the fixed mesh, we need to determine the time period Δt_M so that the interface between the magma and rock just moves a distance being equal to ΔL_M in the interface movement direction. Since the magma is fully solidified within the region of the solidification thickness ΔL_M, the delta function used in Eq. (7.3) should have a value of unity. Except in this solidification region, the delta function should have a value of zero, meaning that there is no magma solidification taking place in other regions. This is the reason why the delta function only has values of unity and zero in the present numerical algorithm. Generally, for the same magma solidification thickness ΔL_M, the solidification time period can vary significantly as the magma solidification proceeds. This requires us to develop a numerical algorithm, in which the finite element mesh is fixed but the integration time-step being equivalent to the solidification time period must be variable.

Based on the above considerations, Eqs. (7.1), (7.2), (7.3), (7.4) and (7.5) can be represented by the following equations:

$$(\rho_R c_{pR})\frac{\partial T_R}{\partial t} = \lambda_R \left(\frac{\partial^2 T_R}{\partial x^2} + \frac{\partial^2 T_R}{\partial y^2}\right) + \delta(x-x_I, y-y_I)f(x, y, t) \quad (x, y) \in V_R-V_M, \tag{7.6}$$

$$\frac{\partial C_T}{\partial t} = D\left(\frac{\partial^2 C_T}{\partial x^2} + \frac{\partial^2 C_T}{\partial y^2}\right) + \delta(x-x_I, y-y_I)Q(x, y, t) \quad (x, y) \in V_R-V_M, \tag{7.7}$$

where δ is the delta function with values equal to unity and zero; $f(x, y, t)$ is the physically equivalent heat source due to the solidification of the intruded magma.

$$f(x, y, t) = \frac{\rho_M\left[L + c_{pM}(T_{IM} - T_m)\right]\left(n_x \frac{\partial x_I}{\partial t} + n_y \frac{\partial y_I}{\partial t}\right)}{\Delta L_{Mk}}, \tag{7.8}$$

where ΔL_{Mk} is the solidification thickness of the intruded magma during a time period Δt_{Mk}; k is the time-step index of integration in the finite element analysis.

The mass source of the released volatile fluids in Eq. (7.7) during the intruded magma solidification can be determined using the solubility of H_2O in the $NaAlSi_3O_8$ melt (Burnham 1979, Barns 1997). In this regard, the mole fraction of H_2O in the $NaAlSi_3O_8$ melt is expressed as follows:

$$X_w^m = \frac{1}{\sqrt{k_w^{mf}}} \quad (X_w^m \le 0.5), \tag{7.9}$$

$$X_w^m = 0.5 + \cfrac{1}{6.25 - \cfrac{2667}{T}} \ln\left(\frac{4}{k_w^{mf}}\right) \qquad (X_w^m > 0.5) \qquad (7.10)$$

where X_w^m is the mole fraction of H_2O in the $NaAlSi_3O_8$ melt; T is in Kelvin; k_w^{mf} is the equilibrium constant for H_2O in melts of feldspar composition.

$$\begin{aligned}\ln k_w^{mf} =\ & 5 + (\ln P)(4.481 \times 10^{-8}T^2 - 1.51 \times 10^{-4}T - 1.137) \\ & + (\ln P)^2(1.831 \times 10^{-8}T^2 - 4.882 \times 10^{-5}T + 4.656 \times 10^{-2}) \\ & + 7.8 \times 10^{-3}(\ln P)^3 - 5.012 \times 10^{-4}(\ln P)^4 + 4.754 \times 10^{-3}T \\ & - 1.621 \times 10^{-6}T^2,\end{aligned} \qquad (7.11)$$

where P is the pressure of the intruded magma; P and T are in bars and Kelvin, respectively.

Using the concept of molar mass, the mass source of the volatile fluids released during solidification from the intruded magma can be expressed as

$$Q(x, y, t) = \frac{X_w^m W_w^m}{\left(\dfrac{X_w^m W_w^m}{\rho_w^m} + \dfrac{(1 - X_w^m)W_{albite}^m}{\rho_{albite}^m}\right)\Delta t_{Mk}}, \qquad (7.12)$$

where Δt_{Mk} is the time period required to complete the magma solidification within a given solidification thickness ΔL_{Mk}; W_w^m and ρ_w^m are the molar mass and density of the volatile fluids in the magma; W_{albite}^m and ρ_{albite}^m are the molecular mass and density of the albite ($NaAlSi_3O_8$) melt. It is noted that using the definition in Eq. (7.12), the mass source of the released volatile fluids has units of the density of the albite ($NaAlSi_3O_8$) melt divided by time.

In the case of the intruded magma temperature being equal to the solidification magma temperature, Eq. (7.8) can be rewritten as

$$f(x, y, t) = \frac{\rho_M L \left(n_x \dfrac{\partial x_I}{\partial t} + n_y \dfrac{\partial y_I}{\partial t}\right)}{\Delta L_{Mk}}. \qquad (7.13)$$

7.2 Implementation of the Equivalent Source Algorithm in the Finite Element Analysis with Fixed Meshes

If the physically equivalent heat source term, $f(x, y, t)$, is determined either analytically or experimentally, Eqs. (7.6) and (7.7) can be directly solved using the conventional finite element method (Zienkiewicz 1977). For dike-like and sill-like intruded magmas, the physically equivalent heat source due to the solidification can

7.2 Implementation of the Equivalent Source Algorithm

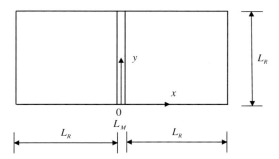

Fig. 7.2 A representative model for the magma solidification problem

be determined analytically through the following ideal experiment. As shown in Fig. 7.2, the sample in the ideal experiment is a rectangular domain filled with the intruding magma in the middle and the intruded rocks on the either side. All the external boundaries of the sample are insulated and the temperatures of the rock and intruded magma are T_R and T_m, respectively. The characteristic length of the rock is L_R, while the characteristic length of the intruded magma is $L_M/2$. It is assumed that the characteristic length of the rock is much larger than that of the intruded magma, which is true for sills and dikes in geology (Johnson and Pollard, 1973, Lister and Kerr 1991, Rubin 1995, Weinberg 1996). Under this assumption, the solidification problem in the ideal experiment can be treated as a one-dimensional Stefan solidification problem, for which the analytical solution is already available (Carslaw and Jaeger 1959).

If the characteristic length of the intruded magma is divided into K equal parts, which are modelled by K finite element meshes of equal length in the magma solidification direction, then the length of the finite element mesh in this direction is $\Delta x = L_M/(2K)$. In order to use the fixed finite element mesh, it is useful to keep Δx constant. For this purpose, we define that the solidification thickness of the intruded magma during a time period Δt_{Mk} is equal to the length of the finite element mesh in the magma solidification direction and, therefore, $\Delta L_{Mk} = \Delta x = L_M/(2K)$. This implies that for this ideal experiment, the scalar product of (n_x, n_y) and $(\partial x_I/\partial t, \partial y_I/\partial t)$ can be expressed as follows:

$$n_x \frac{\partial x_I}{\partial t} + n_y \frac{\partial y_I}{\partial t} = n_x \frac{\partial x_I}{\partial t} \approx \frac{\Delta x}{\Delta t_{Mk}} = \frac{\Delta L_{Mk}}{\Delta t_{Mk}}, \quad (7.14)$$

where Δt_{Mk} is the time period when the solidification boundary moves from one side of the kth finite element in the magma solidification direction at time $t = t_{Mk-1}$ to another side of the kth finite element in the magma solidification direction at time $t = t_{Mk}$.

$$\Delta t_{Mk} = t_{Mk} - t_{Mk-1} \quad (k = 1, 2, 3, \ldots, K). \quad (7.15)$$

The time, at which the intruded magma solidification boundary reaches both sides of the *k*th finite element boundary in the magma solidification direction, can be determined from the analytical solution (Carslaw and Jaeger 1959).

$$t_{Mk} = \frac{k^2(\Delta x)^2}{4\alpha\beta^2}, \quad t_{Mk-1} = \frac{(k-1)^2(\Delta x)^2}{4\alpha\beta^2} \quad (k = 1, 2, 3, \ldots, K), \tag{7.16}$$

where $\alpha = \lambda_R/(\rho_R c_{pR})$ is the thermal diffusivity of the rock; β can be determined from the following transcendental equation:

$$\frac{e^{-\beta^2}}{\beta[1 + erf(\beta)]} = \frac{L\sqrt{\pi}}{c_{pR}(T_m - T_{R0})}, \tag{7.17}$$

where T_{R0} is the initial temperature of the rock; $erf(\beta)$ is the error function of variable β.

$$erf(\beta) = \frac{2}{\sqrt{\pi}} \int_0^\beta e^{-r^2} dr. \tag{7.18}$$

Substituting Eqs. (7.14) into Eq. (7.13) yields the following equation:

$$f(x, y, t_{Mk}) = \frac{\rho_M L}{\Delta t_{Mk}} \quad (k = 1, 2, 3, \ldots, K). \tag{7.19}$$

Finally, substituting Eq. (7.16) into Eq. (7.19) yields the following equation:

$$f(x, y, t_{Mk}) = \frac{4\rho_M L\alpha\beta^2}{(2k-1)(\Delta x)^2} \quad (k = 1, 2, 3, \ldots, K), \tag{7.20}$$

where $\Delta x = \Delta L_{Mk} = L_M/(2K)$ and ΔL_{Mk} is the constant solidification thickness of the intruded magma during the variable time period Δt_{Mk}.

It needs to be pointed out that since the magma solidification boundary passes only through the *k*th finite element from the initial intruded interface between the rock and magma during the time period Δt_{Mk}, both the mass source of the released volatile fluids and the physically equivalent heat source expressed in Eqs. (7.12) and (7.20) only need to be applied in the *k*th finite element, which is numbered from the initial intruded interface between the rock and magma, in the finite element analysis. This means that if a fixed finite element mesh is employed, the variable time step expressed by Eqs. (7.15) and (7.16) needs to be used during the magma solidification period in the finite element analysis. From the numerical analysis point of view, a change in the time step is much easier to implement than a change in the finite element mesh in the finite element analysis. This is the main advantage in using the proposed equivalent source algorithm to simulate the chemical effect of the intruded magma solidification in geological systems.

7.3 Verification and Application of the Equivalent Source Algorithm

Due to the geometrical symmetry of the ideal experiment, analytical solutions for the temperature distribution during the solidification of the intruded magma can be expressed as follows.

$$T_R = T_m \quad \left(0 \le x \le \frac{L_M}{2} - 2\beta\sqrt{\alpha t},\ t \le t_{MMax}\right), \tag{7.21}$$

$$T_R = T_{R0} + \frac{(T_m - T_{R0})\,erfc\left(\dfrac{2x + L_M - 4\beta\sqrt{\alpha t}}{4\sqrt{\alpha t}}\right)}{1 + erf(\beta)} \tag{7.22}$$

$$\left(\frac{L_M}{2} - 2\beta\sqrt{\alpha t} \le x \le L_R,\ t \le t_{MMax}\right),$$

$$T_R = T_{R0} + \frac{\rho_M L_M\left[c_{pM}(T_m - T_{R0}) + L\right]}{2\rho_R c_{pR}\sqrt{\alpha \pi t}} e^{-\frac{x^2}{4\alpha t}} \quad (t > t_{MMax}), \tag{7.23}$$

where $erfc(x) = 1 - erf(x)$ is the complementary error function of variable x; $t_{MMax} = L_M^2/(16\alpha\beta^2)$ is the maximum time instant to complete the solidification of the intruded magma.

In summary, the proposed equivalent algorithm for simulating the chemical effect of the intruded magma solidification includes the following five main steps: (1) For the given values of the initial temperature of the rock, T_{R0}, and the solidification temperature of the intruded magma, T_m, Eq. (7.17) is solved to determine the value of β; (2) Substituting the value of β into Eq. (7.20) yields the value of the physically equivalent heat source due to the solidification of the dike-like and sill-like intruded magma; (3) Eqs. (7.12) and (7.15) are used to calculate the mass source of the released volatile fluids during the intruded magma solidification; (4) The integration time step is determined using Eqs. (7.15) and (7.16); (5) The conventional finite element method is used to solve Eqs. (7.6) and (7.7) for the temperature and concentration distribution in the whole domain during and after solidification of the intruded magma. Note that both the mass source and the physically equivalent heat source are only applied to some elements in the finite element analysis during the solidification process of the intruded magma. The automatic transition from solidification to post-solidification of the intruded magma is another advantage in using the proposed equivalent source algorithm for simulating the chemical effect of the intruded magma solidification in the conventional finite element analysis.

7.3 Verification and Application of the Equivalent Source Algorithm

Since the key part of the proposed algorithm for simulating the chemical effect of the intruded magma during solidification is to transform the original moving interface (i.e. the solidification interface between the rock and intruded magma) problem into

164   7 Simulating Thermal and Chemical Effects of Intruded Magma Solidification Problems

a physical equivalent problem which can be solved using the conventional fixed-mesh finite element method, it is necessary to use a benchmark magma solidification problem, for which the analytical solution is available, to verify the proposed algorithm. As mentioned in the previous section, the effect of the released volatile fluids on the heat transfer and magma solidification is negligible due to the relatively small amount of their volumetric mass sources. On the other hand, transport of the released volatile fluids can be treated as a conventional mass transport problem, for which the conventional finite element method has been used for many years (Zienkiewicz 1977). This indicates that we only need to verify the proposed algorithm for simulating the heat transfer process associated with the intruded magma solidification. For the above reasons, only the heat transfer process associated with the benchmark magma solidification problem is considered to verify the proposed algorithm. Figure 7.3 shows the finite element mesh of the benchmark magma solidification problem. In this figure, the original magma intruded region, which is indicated in black in the finite element mesh of the whole computational domain, is

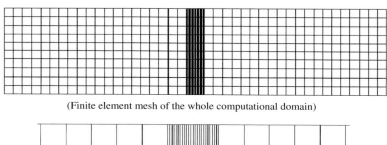

(Finite element mesh of the whole computational domain)

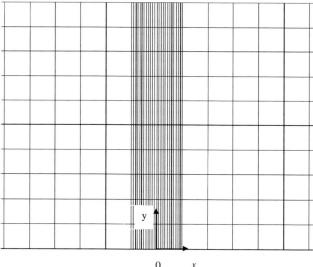

(Detailed mesh for the intruded magma solidification domain)

Fig. 7.3 Finite element mesh of the benchmark magma solidification problem

7.3 Verification and Application of the Equivalent Source Algorithm

modeled using 240 four-node quadrilateral finite elements, while the whole computational domain is modeled with 640 four-node quadrilateral finite elements. The length and width of the whole computational domain are 42 m and 20 m, respectively. The width of the intruded dike-like magma is assumed to be 2 m in this benchmark problem. For the purpose of comparing the numerical solutions with the corresponding analytical ones, the thermal properties of the intruded magma are assumed to be the same as those of the surrounding rocks. The following parameters are used in the finite element analysis: the densities of both the magma and surrounding rocks are 2900 kg/m^3; specific heat is 1200 J/(kg×°C); thermal conductivity is 1.74 W/(m×°C); the latent heat of fusion of the intruded magma is 3.2×10^5 J/kg. Since the temperature difference between the intruded magma and the surrounding rocks is an important indicator of this benchmark problem, it is assumed to be 1000°C in the numerical computation.

Using the above thermal properties and Eq. (7.17), the value of β is determined to be 0.73 approximately. In order to examine the effect of time-steps of the intruded magma solidification on the overall accuracy of the numerical solution, two computational models, namely a one-step solidification model and a three-step solidification model, are considered in the transient finite element analysis. For the one-step solidification model, the intruded magma is solidified just in one time step, which is 10.86 days from Eq. (7.16). The corresponding physically-equivalent heat source determined from Eq. (7.20) is 989.06 W/m^3 for this one-step solidification model. However, for the three-step solidification model, the intruded magma is solidified in three variable time steps so that the intruded magma solidified equal distance from the beginning interface between the magma and surrounding/solidified rocks. Using Eq. (7.16), the three variable time steps are $10.86/9$, $10.86/3$ and $(5 \times 10.86)/9$ days and the corresponding physically-equivalent heat sources are 8901.54 W/m^3, 2967.18 W/m^3 and 1780.31 W/m^3 respectively. After completion of the intruded magma solidification, the constant time step of 10.86 days is used throughout the rest of the transient finite element analysis.

Figures 7.4 and 7.5 show the comparisons of the analytical solutions with the corresponding numerical solutions from the one-step solidification model and the three-step solidification model respectively. It is obvious that the numerical solutions of the temperature difference between the intruded magma and the surrounding rocks, which are obtained from both the one-step solidification model and the three-step solidification model, have very good agreement with the corresponding analytical ones. For example, at the time instant of 21.73 days (i.e. $t = 21.73$ days), the numerical solutions of the maximum temperature difference between the intruded magma and the surrounding rocks are 732.1°C and 733.5°C for the one-step solidification model and the three-step solidification model respectively, while the corresponding analytical solution of the maximum temperature difference between the intruded magma and the surrounding rocks is 737.8°C. The relative numerical error between the numerical and analytical solutions for the maximum temperature difference between the intruded magma and the surrounding rocks is 0.78% and 0.58% for the one-step solidification model and the three-step solidification model respectively. This demonstrates that the numerical model based on

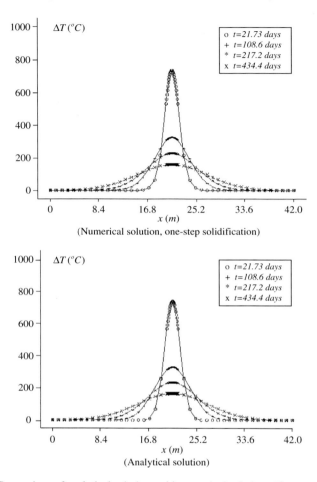

Fig. 7.4 Comparison of analytical solutions with numerical solutions (One-step solidification)

the proposed equivalent algorithm for simulating solidification effects of magma intrusion problems in porous rocks can produce highly accurate numerical solutions. Since the numerical solution from the one-step solidification model has similar accuracy to that from the three-step solidification model, the one-step solidification model is useful for simulating thermal effects of dike-like magma intrusion problems in porous rocks. For the purpose of examining the numerical solution sensitivity to the mesh density in the magma solidification direction, we have doubled and tripled the mesh density in the magma solidification region. Since the solidification interface between the rock and intruded magma is precisely considered in the magma solidification direction, all the three meshes of the original mesh density (i.e. 240 finite elements to simulate the intruded magma in the solidification direction), the doubled mesh density (i.e. 480 finite elements to simulate the intruded magma in the solidification direction) and the tripled mesh density (i.e. 720 finite

7.3 Verification and Application of the Equivalent Source Algorithm

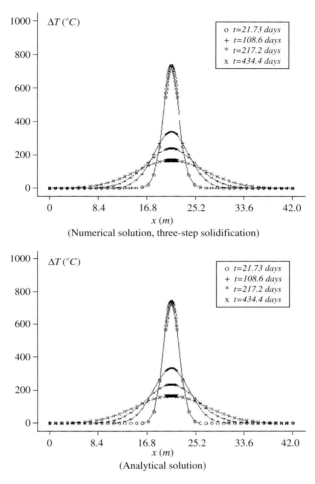

Fig. 7.5 Comparison of analytical solutions with numerical solutions (Three-step solidification)

elements to simulate the intruded magma in the solidification direction) produce similar numerical results. This indicates that, due to the precise consideration of the magma solidification interface, the present numerical algorithm is not sensitive to the mesh density in the solidification direction of the magma solidification region.

It is noted that, for intruded magma dikes and sills of large aspect ratios, one could have expected on dimensional considerations that the solidification of the intruded magma could be treated by means of a one-dimensional model in the solidification direction and that such an approximation is asymptotically valid to the leading order in the aspect ratio of the dike and sill. Such a one-dimensional (1D) solidification model could be interactively used with two-dimensional (2D) models for simulating the thermal field in the surrounding rocks. The use of the resulting hybrid 1D-2D model may avoid the need to introduce equivalent volumetric heat and mass source terms. However, since the 1D and 2D models are considered separately and interactively, one must use the 1D model to determine the temperature and thermal flux boundary conditions on the solidification interface between the rock and intruded magma, and then apply such boundary conditions to the 2D model for simulating the thermal field in the surrounding rocks. Since the magma solidification is time-dependent, the above-mentioned boundary conditions need to be determined and applied in the hybrid 1D-2D model repeatedly for each time step during the intruded magma solidification. This certainly adds numerical modelling complexity during the magma solidification, because the previous boundary conditions must be modified and the current ones must be applied to the 2D model. This complexity has been avoided by using the proposed numerical algorithm in this study, since the magma solidification in the intruded magma region and heat transfer in the surrounding rocks are considered simultaneously, rather than separately, in the finite element analysis. Due to this obvious advantage in using the proposed model, there is no need to develop the hybrid 1D-2D model for simulating the solidification problems associated with intruded magma dikes and sills.

After the verification of the proposed equivalent source algorithm for simulating solidification effects of dike-like magma intrusion problems in porous rocks, it has been applied to investigate the chemical effects of a dike-like magma intrusion/solidification problem in the upper crust of the Earth. Figure 7.6 shows the finite element mesh of the dike-like magma intrusion problem. The length and width of the whole computational domain are 40.2 *km* and 10 *km*, the region of the intruded dike-like magma (as indicated by black colour) is 0.2 *km* and 6 *km* in the horizontal and vertical directions, respectively. The intruded magma region is modelled with 144 four-node quadrilateral finite elements, while the whole computational domain is modelled with 1040 four-node quadrilateral finite elements in the transient finite element analysis. The thermal properties of this application problem are exactly the same as those of the previous benchmark magma solidification problem. The diffusivity of the released volatile fluids is 2×10^{-6} m^2/s. The pressure of the intruded magma is assumed to be 3000 bars. Since the width of the intruded magma for this application problem is 100 times that for the previous benchmark magma solidification problem, the solidification time of the intruded magma for this application problem is 10,000 times that for the previous benchmark magma solidification problem,

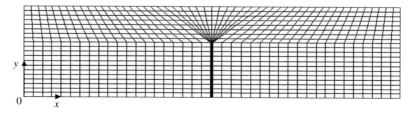

Fig. 7.6 Finite element mesh of the dike-like magma intrusion problem

7.3 Verification and Application of the Equivalent Source Algorithm 169

as indicated from Eq. (7.16). Therefore, the time step of 9.3826×10^9 s (which is approximately equal to 300 years) is used in the transient finite element analysis of the dike-like magma intrusion problem in the upper crust of the Earth. For this application problem, the boundary condition is that the temperature at the top of the computational model is 20°C throughout the transient finite element analysis. The initial conditions are as follows: the initial temperature of the intruded dike-like magma is 1230°C and the initial temperature at the rest (i.e., except for the intruded magma) of the bottom of the computational model is 320°C. The initial average temperature of the rocks surrounding the intruded dike-like magma is 230°C. The initial concentration of the released volatile fluids is assumed to be zero in the whole

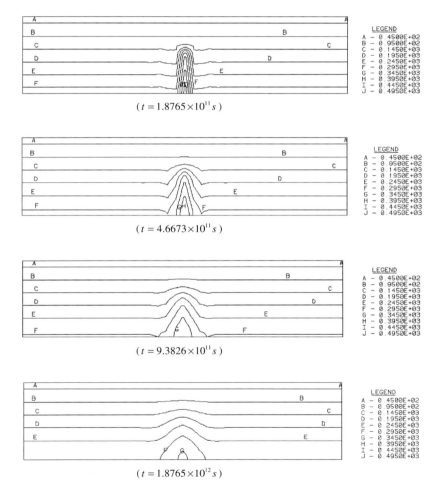

Fig. 7.7 Temperature distributions of the dike-like magma intrusion problem at different time instants

computational domain. Under the above temperature and pressure conditions of the intruded magma, the density of the released volatile fluids is 374.7 kg/m^3 (Haar et al. 1984), while the density of the albite melt is 2700 kg/m^3. In addition, the mole fraction of the released volatile fluids from the albite melt can be determined as 0.5306 from Eqs. (7.10) and (7.11), which results in a mass source of 1.433×10^{-8} kg/(m$^3 \times$ s) for the released volatile fluids in the computation.

Figure 7.7 shows the temperature distributions of the dike-like magma intrusion problem at four different time instants. It is clear that with increasing time, the total temperature localization area generated by the intruded magma becomes larger and

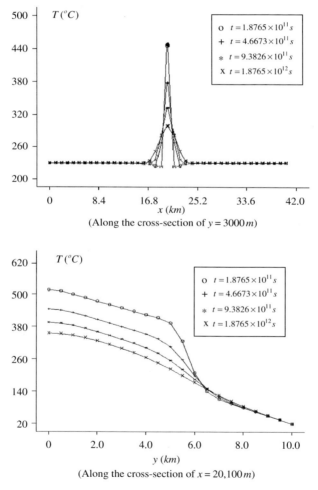

Fig. 7.8 Temperature distributions of the dike-like magma intrusion problem along two different cross-sections

7.3 Verification and Application of the Equivalent Source Algorithm 171

larger, but the maximum temperature generated by the intruded magma becomes smaller and smaller. These phenomena can be clearly seen from Fig. 7.8, where the temperature distributions of the dike-like magma intrusion problem are displayed along two typical cross-sections of the computational model. For instance, the maximum temperature along the cross-section of $x = 20,100$ m is 514.8°C, 444.4°C, 394.4°C and 354.9°C for the time instants of 1.8765×10^{11} s, 4.6673×10^{11} s, 9.3826×10^{11} s and 1.8765×10^{12} s respectively. Although the total temperature localization area generated by the intruded magma is limited in the top part of the

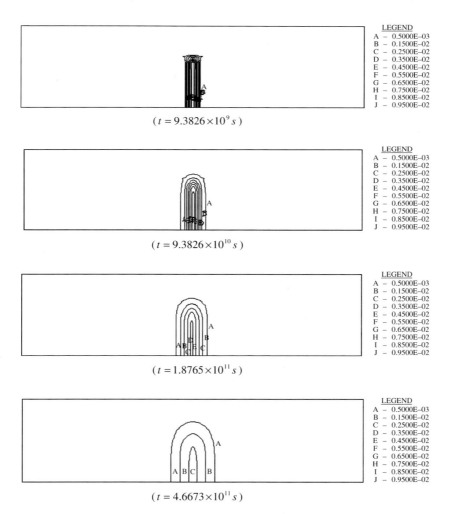

Fig. 7.9 Concentration distributions of volatile fluids for the dike-like magma intrusion problem at different time instants (Whole system view)

computational model, it is large enough to cause convective pore-fluid flow within the whole system (Zhao et al. 1997a), if the porous rocks are saturated by the pore-fluid in the upper crust of the Earth. Once this convective pore-fluid flow takes place, temperature localization in the top part of the computational model can be significantly enhanced so that a favourable region for ore body formation and mineralization can be created just above the intruded dike-like magma (Zhao et al. 1998a).

Figure 7.9 shows the concentration distributions of the volatile fluids for the whole system of the dike-like magma intrusion problem at four different times, while Figs. 7.10 and 7.11 show the detailed concentration distributions of the volatile fluids for a zoomed-in part of the dike-like magma intrusion problem at the same four different times. It is clear that with an increase in time, the total concentration area of the volatile fluids generated during the intruded magma solidification becomes larger and larger, but the maximum concentration of the volatile fluids generated by the intruded magma becomes smaller and smaller. These phenomena can be clearly seen from Figs. 7.10 and 7.11, where the concentration distributions of

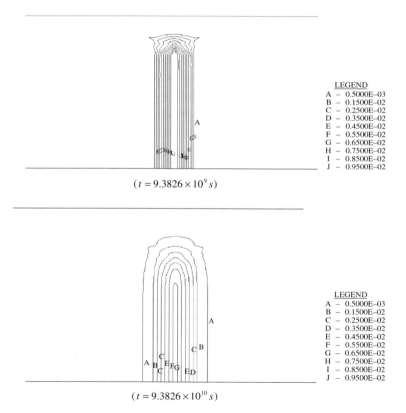

Fig. 7.10 Concentration distributions of volatile fluids for the dike-like magma intrusion problem at different time instants (Zoomed-in view)

7.3 Verification and Application of the Equivalent Source Algorithm

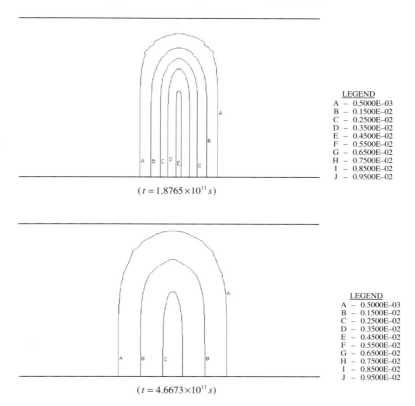

Fig. 7.11 Concentration distributions of volatile fluids for the dike-like magma intrusion problem at different time instants (Zoomed-in view)

the volatile fluids during the dike-like magma solidification are displayed within the zoomed-in part of the computational model. Since the released volatile fluids can react chemically with the surrounding rocks (Zhao et al. 2001d), an aureole, which can be clearly observed in these figures, is formed during the intruded magma solidification. Although the total area of the aureole generated by the intruded magma is limited in the computational model, it is large enough to cause ore body formation and mineralization to take place within the system. This indicates that the released volatile fluids can create a favourable region for ore body formation and mineralization to take place through some chemical reactions around the intruded dike-like magma.

Chapter 8
The Particle Simulation Method for Dealing with Spontaneous Crack Generation Problems in Large-Scale Geological Systems

Cracking and fracturing are one class of major failure mechanisms in brittle and semi-brittle materials. Crustal materials of the Earth can be largely considered as brittle rocks, and so cracking and fracturing phenomena are ubiquitous. Cracks created within the Earth's crust often provide a very useful channel for mineral-bearing fluids to flow, particularly from the deep crust into the shallow crust of the Earth. If other conditions such as fluid chemistry, mineralogy, temperature and pressure are appropriate, ore body formation and mineralization can take place as a result of such fluid flow. Because of the ever-increasing demand for mineral resources in the contemporary world, exploration for new mineral resources has become one of the highest priorities for many industrial countries. For this reason, extensive studies (Garven and Freeze 1984, Yeh and Tripathi 1989, 1991, Steefel and Lasaga 1994, Raffensperger and Garven 1995, Zhao et al. 1997a, Schafer et al. 1998a, b, Zhao et al. 1998a, Xu et al. 1999, Zhao et al. 2000b, Schaubs and Zhao 2002, Zhao et al. 2002a, 2003e, 2005a) have been conducted to understand the detailed physical and chemical processes that control ore body formation and mineralization within the upper crust of the Earth. Thus, the numerical simulation of spontaneous crack generation in brittle rocks within the upper crust of the Earth has become an important research topic in the field of computational geoscience.

The numerical simulation of crack initiation has existed from the inception and development of numerical fracture mechanics. This was the direct outcome of combining conventional fracture mechanics with numerical methods such as the finite element and the boundary element methods. Due to the increased capability of numerical fracture mechanics for considering complicated geometry and boundary conditions, it significantly extends the applicability of conventional fracture mechanics to a wide range of practical problems in civil and geotechnical engineering fields. In the study of numerical simulation of crack initiation and evolution, it is usually assumed that a crack initiates when the value of a local principal stress attains the tensile strength of the brittle material. Once a crack is initiated, propagation of the crack, including the propagation direction and incremental growing length of the crack, can be determined by crack propagation theories established in fracture mechanics. Although there are at least three kinds of mixed-mode crack propagation theories available, namely a theory based on propagation driven by the maximum circumferential tensile stress (Erdogan and Sih 1963), a theory based on propagation driven by strain energy density near the crack tip (Sih and Macdonald 1974) and a theory based on crack propagation driven by maximum energy release rate (Rice 1968, Hellen 1975, Lorenzi 1985, Li et al. 1985), the maximum circumferential tensile propagation theory has been widely used in the finite element analysis of crack initiation and propagation problems because it is not only physically meaningful, but it is also easily implemented in a finite element code. It has been widely demonstrated that, for the finite element simulation of a two-dimensional fracture mechanics problem, the triangular quarter-point element, which is formed by collapsing one side of a quadrilateral 8-noded isoparametric element, results in improved numerical results for stress intensity factors near a crack tip (Barsoum 1976, 1977). Similarly, for the finite element modelling of a three-dimensional fracture mechanics problem, the prismatic quarter-point element, which is formed by collapsing one face of a cubic 20-noded isoparametric element, can also lead to much better numerical results for stress intensity factors near a crack tip (Barsoum 1976, 1977, Ingraffea and Manu 1980). This kind of quarter-point element can be used to simulate the $1/\sqrt{r}$ singularity in elastic fracture mechanics by simply assigning the same displacement at the nodes located on the collapsed side or face of the element, while it can be also used to simulate the $1/r$ singularity in perfect plasticity by simply allowing different displacements at the nodes located on the collapsed side or face of the element. In order to simulate appropriately displacement discontinuities around a crack, automatic meshing and re-meshing algorithms (Zienkiewicz and Zhu 1991, Lee and Bathe 1994, Khoei and Lewis 1999, Kwak et al. 2002, Bouchard et al. 2003) have been developed in recent years. However, such finite element methods with automatic meshing and re-meshing algorithms have been applied for solving crack generation and propagation problems in systems of only a few cracks.

For the purpose of removing this limitation associated with the conventional numerical method (which is usually based on continuum mechanics) in simulating a large number of spontaneously generated cracks, particle simulation methods, such as the distinct element method developed as a particle flow code (Cundall and Strack 1979, Cundall 2001, Itasca Consulting Group, inc. 1999, Potyondy and Cundall 2004), provides a very useful tool to deal with this particular kind of problem. Since displacement discontinuities at a contact between two particles can be readily considered in these methods, the formulation based on discrete particle simulation is conceptually simpler than that based on continuum mechanics, because crack generation at a contact between two particles is a natural part of the particle simulation process.

Even though the particle simulation method was initially developed for solving soil/rock mechanics, geotechnical and other engineering problems (Cundall and Strack 1979, Bardet and Proubet 1992, Thomton et al. 1999, Tomas et al. 1999, Salman and Gorham 2000, Klerck et al. 2004, Owen et al. 2004, McBride et al. 2004 and Schubert et al. 2005, Zhao et al. 2006f), it has been used to deal with a large number of geological and geophysical problems in both two and three dimensions (Saltzer and Pollard 1992, Antonellini and Pollard 1995, Donze et al. 1996, Scott 1996, Strayer and Huddleston 1997, Camborde et al. 2000, Iwashita and Oda 2000, Burbidge and Braun 2002, Strayer and Suppe 2002, Finch et al. 2003, 2004, Imber et al. 2004, Zhao et al. 2007b, c, d, Zhao et al. 2008g). Although the particle simulation method has been successfully used to solve these large-scale geological problems, little work, if any, has been reported on using the particle simulation method to deal with spontaneous crack generation problems in the upper crust of the Earth. Nevertheless, the particle simulation method has been developed to simulate microscopic crack generation in small-scale laboratory specimens and mining sites (Itasca Consulting Group, inc. 1999, Potyondy and Cundall 2004). Since both the time-scale and the length-scale are quite different between laboratory specimens and geological systems, it is necessary to deal with an upscale issue when the particle simulation method is applied to solve spontaneous crack generation problems in the upper crust of the Earth. Due to the relative slowness of some geological processes, many geological systems can be treated as quasi-static ones (that is, inertia is neglected), at least from the mathematical point of view. Because the mechanical response of a quasi-static system is theoretically independent of time, the time-scale issue can be eliminated in the particle simulation of a quasi-static system. For this reason, this chapter is restricted to deal with the particle simulation of spontaneous crack generation problems within large-scale quasi-static geological systems.

It is however computationally prohibitive to simulate a whole geological system using real physical particles of a microscopic length-scale, even though with modern computer capability. To overcome this difficulty, the following three approaches are often used in dealing with the particle simulation of geological length-scale problems. In the first approach, a geological length-scale problem, which is usually of a kilometer-scale, is scaled down to a similar problem of a small length-scale (i.e. a meter-scale or a centimeter-scale) and then particles of small length-scale (e.g. a microscopic length-scale) are used to simulate this small length-scale geological problem (e.g. Antonellini and Pollard 1995, Imber et al. 2004, Schopfer et al. 2006). In the second approach, the geological length-scale problem is directly simulated using particles with a relative large length-scale (e.g. Strayer and Huddleston 1997, Burbidge and Braun 2002, Strayer and Suppe 2002, Finch et al. 2003, 2004). In this case, the particles used in the simulation can be considered as the representation of a large rock block. Although detailed microscopic deformation cannot be simulated using the second approach, the macroscopic deformation pattern can be reasonably simulated in this approach (e.g. Strayer and Huddleston 1997, Burbidge and Braun 2002, Strayer and Suppe 2002, Finch et al. 2003, 2004). This implies that if the macroscopic deformation process of a geological length-scale system is of interest, then the second approach can produce useful simulation results. This is particularly true for understanding the controlling process of ore body formation and mineralization, in which the macroscopic length-scale geological structures are always of special interest. In the third approach, the combined use of both a block-mechanics-based particle simulation method and a continuum-mechanics-based method, such as the finite element method (Zienkiewicz 1977, Lewis and

Schrefler 1998) or the finite difference method, are used to simulate the whole geological system (Potyondy and Cundall 2004, Suiker and Fleck 2004, Fleck and Willis 2004). In this approach, the continuous deformation range of a system is simulated using the continuum-mechanics-based method, and the discontinuous deformation range of the system is simulated using the block-mechanics-based particle simulation method. Since large element sizes can be used to simulate the continuous deformation range for a quasi-static geological system, the third approach is computationally more efficient than the first and second approaches in dealing with large systems. However, the computational difficulty associated with the third approach is that an adaptive interface between the continuous deformation range and the discontinuous deformation range must be appropriately developed. This means that if the discontinuous deformation range is obvious *a priori* for some kind of geological problem, then the third approach is computationally very efficient. Otherwise, the efficiency of using the third approach might be greatly reduced due to the ambiguity in defining interfaces between the continuous and the discontinuous deformation ranges. For this reason, the first and second approaches are commonly used to simulate large-scale geological systems using the particle simulation method.

Because of the wide use of both the first and the second approaches, the following scientific questions are inevitably posed: Are two particle simulation results obtained from both the first and the second approaches consistent with each other for the same geological problem? If the two particle simulation results are consistent, then what is the intrinsic relationship between the two similar particle models used in the first approach and the second approach? To the best knowledge of the authors, these questions remain unanswered. Therefore, we will develop an upscale theory associated with the particle simulation of two-dimensional quasi-static geological systems at different length-scales to clearly answer these two scientific questions. The present upscale theory is of significant theoretical value in the particle simulation of two-dimensional systems, at least from the following two points of view. (1) If the mechanical response of a particle model of a small length-scale is used to investigate indirectly that of a large length-scale, then the present upscale theory provides the necessary conditions that the particle model of the small length-scale needs to satisfy so that similarity between the mechanical responses of the two different length-scale particle models can be maintained. (2) If a particle model of a large length-scale is used to investigate directly the mechanical response of the model, then the present upscale theory can be used to determine the necessary particle-scale mechanical properties from the macroscopic mechanical properties that are obtained from either a laboratory test or an *in-situ* measurement. Because the particle simulation method has been used to solve many kinds of scientific and engineering problems, the present upscale theory can be directly used to extend the application range of the particle simulation results for geometrically-similar problems without a need to conduct another particle simulation. This means that the present upscale theory compliments existing particle simulation method, and is applicable for practical applications in many scientific and engineering fields.

8.1 Basic Formulations of the Particle Simulation Method

The basic idea behind the particle simulation method is that either a granular material or a solid material can be simulated using an assembly of particles. In the case of simulating granular materials, there is no cementation between two adjacent particles so that the particles are not bonded together in the assembly. However, in the case of simulating solid materials, particles are bonded together through cementation. Although particles may have different shapes and sizes, they are assumed to be rigid. Thus, the motion of a particle can be described using the motion of its mass center. A small overlapping between two particles is allowed so that deformation of the particle assembly can be simulated. The magnitude of the overlapping depends on both the contact force and the stiffness of a contact between any two particles. This means that a relationship between contact force and displacement needs to be used to calculate the contact force at a contact between two particles. In the case of simulating solid materials such as brittle rocks, the strength of the material can be simulated by using the strength of the bond at a contact between two particles. Once the contact force reaches or exceeds the strength of the bond at a contact between two particles, the bond is broken so that a microscopic crack is created to represent the failure of the material at this particular bond. This consideration is very useful for explicitly simulating the initiation of spontaneous and random microscopic cracks in a brittle material, because there is no need to artificially describe any microscopic flaws and cracks in the beginning of a numerical simulation. Although the above idea was initially proposed in the discrete element method (Cundall and Strack 1979, Cundall 2001, Itasca Consulting Group, inc. 1999), it has been recently enhanced and built into a two-dimensional Particle Flow Code. In this regard, the particle simulation method can be viewed as a particular kind of discrete element method.

Considering a particle (i.e. particle α) shown in Fig. 8.1, the position of the particle is described using its x and y coordinates in the coordinate system shown. According to Newton's second law, the motion of the particle can be represented using the motion of its mass center as follows:

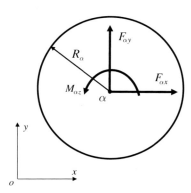

Fig. 8.1 Definition of the motion of a particle

$$F_{\alpha x} = m_\alpha \frac{d^2 x_\alpha}{dt^2}, \quad (8.1)$$

$$F_{\alpha y} = m_\alpha \frac{d^2 y_\alpha}{dt^2}, \quad (8.2)$$

$$M_{\alpha z} = I_\alpha \frac{d^2 \theta_{\alpha z}}{dt^2}, \quad (8.3)$$

where $F_{\alpha x}$, $F_{\alpha y}$ and $M_{\alpha z}$ are the total translational force components and rotational moment exerted on the mass center of particle α; m_α and I_α are the mass and principal moment of inertia with respect to the z axis that is perpendicular to the x-y plane; x_α and y_α are the horizontal and vertical coordinates expressing the position of particle α; $\theta_{\alpha z}$ is the rotation angle of particle α with respect to the principal rotational axis of the particle.

For a two-dimensional disk-shaped particle, the corresponding principal moment of inertia is expressed as follows:

$$I_\alpha = \frac{1}{2} m_\alpha R_\alpha^2, \quad (8.4)$$

where R_α is the radius of particle α.

The translational forces and rotational moment expressed in the above equations can be calculated by adding all the forces and moments exerted on the particle. It is noted that a particle can contact several particles at the same time so that there are several contact forces exerted on the particle. For a particular contact (i.e. contact C) between two particles (i.e. particles A and B) shown in Fig. 8.2, the normal component of the contact force can be calculated using the following formula:

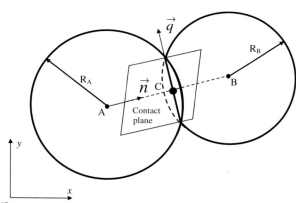

Fig. 8.2 Definition of the contact between two particles

8.1 Basic Formulations of the Particle Simulation Method

$$F^n = \begin{cases} k_n u_n & (u_n \geq \bar{u}_n) \\ 0 & (u_n < \bar{u}_n) \end{cases} \quad \text{(Before the normal contact bond is broken),} \tag{8.5}$$

$$F^n = \begin{cases} k_n u_n & (u_n \geq 0) \\ 0 & (u_n < 0) \end{cases} \quad \text{(After the normal contact bond is broken),} \tag{8.6}$$

where k_r is the stiffness of contact C; u_n is the normal displacement at contact C; $\bar{u}_n$ is the critical normal displacement in correspondence with the normal contact bond breakage at contact C; F^n is the normal contact force at contact C. The contact force at contact C is assumed to be exerted from particle A on particle B.

$$u_n = R_A + R_B - \sqrt{(x_B - x_A)^2 + (y_B - y_A)^2}, \tag{8.7}$$

where R_A and R_B are the radii of particles A and B; x_A and y_A are the horizontal and vertical coordinates expressing the position of particle A; x_B and y_B are the horizontal and vertical coordinates expressing the position of particle B.

Using the definition expressed in Fig. 8.2, the position of contact C can be described as

$$x_C = x_A + \left(R_A - \frac{1}{2}u_n\right)n_x, \tag{8.8}$$

$$y_C = y_A + \left(R_A - \frac{1}{2}u_n\right)n_y, \tag{8.9}$$

where x_C and y_C are the horizontal and vertical coordinates expressing the position of contact C; n_x and n_y are the direction cosines of the normal vector with respect to the horizontal and vertical axes respectively.

The tangential component of the contact force at contact C can be calculated in an incremental manner as follows:

$$\Delta F^s = \begin{cases} -k_s \Delta u_s = -k_s(V_S \Delta t) & (u_n \geq \bar{u}_n) \\ 0 & (u_n < \bar{u}_n) \end{cases} \quad \begin{array}{l}\text{(before the normal} \\ \text{contact bond is broken),}\end{array} \tag{8.10}$$

$$\Delta F^s = \begin{cases} -k_s \Delta u_s = -k_s(V_S \Delta t) & (u_n \geq 0) \\ 0 & (u_n < 0) \end{cases} \quad \begin{array}{l}\text{(after the normal} \\ \text{contact bond is broken),}\end{array} \tag{8.11}$$

where k_s is the tangential stiffness of contact C; Δu_s is the incremental tangential displacement at contact C; ΔF^s is the corresponding tangential component of the contact force; V_S is the tangential shear velocity at contact C; Δt is the time step in the numerical simulation.

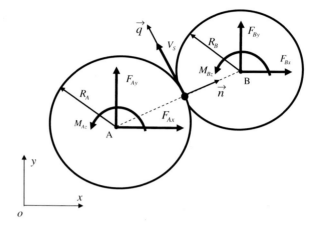

Fig. 8.3 Definition of the shear velocity at a contact between two particles

As shown in Fig. 8.3, the magnitude of the tangential shear velocity, V_S, at contact C can be determined using the relative motion of particles A and B.

$$V_S = -\left(\frac{dx_B}{dt} - \frac{dx_A}{dt}\right)n_y + \left(\frac{dy_B}{dt} - \frac{dy_A}{dt}\right)n_x - \omega_{Bz}\sqrt{(x_B - x_C)^2 + (y_B - y_C)^2}$$
$$- \omega_{Az}\sqrt{(x_A - x_C)^2 + (y_A - y_C)^2}$$
(8.12)

where ω_{Az} and ω_{Bz} are the rotational angular speeds of particles A and B with respect to their rotational axes, which are parallel to the z axis and passing through the corresponding mass centers of each of the two particles respectively.

It is noted that the first two terms in Eq. (8.12) represent the contributions of the relative translational motion to the relative shear velocity between the two particles, whereas the last two terms denote the contributions of the relative rotational motion to the relative shear velocity between the two particles.

For any given time instant, t, the tangential component of the contact force at contact C can be calculated by adding the contact force increment expressed in Eq. (8.11) into the tangential component at the previous time, $t - \Delta t$.

$$F_t^s = F_{t-\Delta t}^s - k_s(V_S \Delta t) \leq \mu F^n, \qquad (8.13)$$

where F_t^s and $F_{t-\Delta t}^s$ are the tangential components of the contact force at t and $t - \Delta t$ respectively; μ is the friction coefficient at contact C. It needs to be pointed out that Eq. (8.13) holds true only when the normal component of the contact force is greater than zero.

The normal and tangential components of the contact force at contact C can be straightforwardly decomposed into the horizontal and vertical components as follows:

8.1 Basic Formulations of the Particle Simulation Method

$$F_{Cx} = F^n n_x + F_t^s q_x, \qquad (8.14)$$

$$F_{Cy} = F^n n_y + F_t^s q_y, \qquad (8.15)$$

where F_{Cx} and F_{Cy} are the horizontal and vertical components of the contact force at contact C; q_x and q_y are the direction cosines of the tangential vector at contact C with respect to the horizontal and vertical axes respectively.

Consequently, the translational and rotational forces exerted on particles A and B due to their contact at point C can be calculated using the following formulas:

$$F_{Ax} = -F_{Cx}, \qquad F_{Ay} = -F_{Cy}, \qquad (8.16)$$

$$F_{Bx} = F_{Cx}, \qquad F_{By} = F_{Cy} \qquad (8.17)$$

$$M_{Az} = -[(x_C - x_A)F_{Cx} + (x_C - x_A)F_{Cy} + (y_C - y_A)F_{Cx} + (y_C - y_A)F_{Cy}], \qquad (8.18)$$

$$M_{Bz} = (x_C - x_B)F_{Cx} + (x_C - x_B)F_{Cy} + (y_C - y_B)F_{Cx} + (y_C - y_B)F_{Cy}, \qquad (8.19)$$

where F_{Ax}, F_{Ay} and M_{Az} are the translational force components and rotational moment exerted on the mass center of particle A; F_{Bx}, F_{By} and M_{Bz} are the translational force components and rotational moment exerted on the mass center of particle B.

Since a particle may have contacts with several particles, it is necessary to search the number of contacts for the particular particle under consideration. Therefore, the total translational forces and rotational moment exerted on a particle can be calculated by adding the contributions of all the contacts to the translational forces and rotational moments exerted on the particle. After the total translational forces and rotational moment are calculated in a particle by particle manner, the central finite difference method is used to solve the equations of motion expressed by Eqs. (8.1), (8.2), and (8.3) for each of the particles in the simulation system, so that new displacements can be determined and the position of each particle can be updated. As a result, a solution loop is formed for each time step in the particle simulation method. This solution loop is comprised of the following four sub-steps: (1) From the position of particles at the beginning of a calculation time step, Eqs. (8.16), (8.17) and (8.19) are used to calculate the contributions of a contact between a particle and each of its surrounding particles to the translational force components and rotational moment exerted on the mass center of both the particle and the surrounding particle; (2) By adding the contributions of all the contacts of a particle to the translational forces and rotational moments exerted on the particle, the total translational forces and rotational moment exerted on the particle are calculated. (3) Sub-steps (1) and (2) are repeated for all the particles so that the total translational forces and rotational moment exerted on every particle in the simulation are calculated. (4) Using the central finite difference method, the equations of motion expressed by Eqs. (8.1), (8.2) and (8.3) are solved for each of the particles in the simulation system, so that new displacements can be determined and the position of every particle can be updated at the end of the calculation time step. The above-mentioned solution loop is repeated for each calculation time step until the final stage of the simulation is reached.

8.2 Some Numerical Simulation Issues Associated with the Particle Simulation Method

Although the particle simulation method such as the distinct element method was developed more than two decades ago (Cundall and Strack 1979), some numerical issues associated with it may need to be further addressed. These include: (1) an issue caused by the difference between an element used in the finite element method and a particle used in the distinct element method; (2) an issue resulting from using the explicit dynamic relaxation method to solve a quasi-static problem; and (3) an issue stemming from an inappropriate loading procedure used in the particle simulation method. Although some aspects of these numerical simulation issues have been briefly discussed (Cundall and Strack 1979, Cundall 2001, Itasca Consulting Group, inc. 1999, Potyondy and Cundall 2004), we will discuss them in greater detail so that their impact on the particle simulation results of spontaneous crack generation problems within large-scale quasi-static systems can be thoroughly understood.

8.2.1 Numerical Simulation Issue Caused by the Difference between an Element and a Particle

In order to investigate the numerical simulation issue associated with the difference between an element used in the finite element method and a particle used in the distinct element method, we need to understand how an element and a particle interact with their neighbors. Figure 8.4 shows the comparison of a typical four-node element used in the finite element method with a typical particle used in the distinct element method. In the finite element method, the degree-of-freedom is represented by the nodal points of the element, while in the distinct element one, the degree-of-freedom is represented by the mass center of the particle. In this regard, the particle may be viewed as a rigid element of only one node. With a two-dimensional material deformation problem taken as an example, the displacement along the common side between two elements is continuous, implying that there is no overlap between any two elements in the conventional finite element method. However, particle overlap is allowed in the distinct element method. Since the element is deformable, the mechanical properties calculated at the nodal point of an element are dependent on the macroscopic mechanical properties of the element so that

8.2 Some Numerical Simulation Issues Associated with the Particle Simulation Method   185

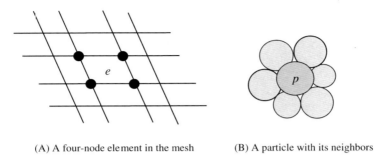

(A) A four-node element in the mesh   (B) A particle with its neighbors

Fig. 8.4 Comparison of a four-node element with a particle

the nodal mechanical property is called the consistent mechanical property of the element. On the other hand, a particle used in the distinct element method is assumed to be rigid and therefore, the mechanical property at a contact between the particle and its neighboring particle is called the lumped mechanical property. Due to this difference, the consistent mechanical properties (i.e. stiffness matrix) of an element are directly calculated from the macroscopic mechanical properties and constitutive law of the element material. This means that once the macroscopic mechanical properties and constitutive law of the element material are available from a laboratory test or a field measurement, the finite element analysis of a deformation problem can be straightforwardly carried out using these macroscopic mechanical properties and constitutive laws of the material. On the contrary, because the particle-scale mechanical properties of materials, such as particle stiffness and bond strength, are used in the distinct element method but are not known *a priori*, it is important to deduce these particle-scale mechanical properties of materials from the related macroscopic ones measured from both laboratory and field experiments. This indicates that an inverse problem needs to be solved through the numerical simulation of a particle system. Thus, the numerical simulation question introduced by the difference between an element used in the continuum-mechanics-based finite element method and a particle used in the discrete-block-mechanics-based distinct element method is how to determine particle-scale mechanical properties from macroscopic mechanical properties available from both laboratory and field experiments. Clearly, the problem associated with this numerical simulation issue cannot be effectively solved unless the required particle-scale mechanical properties of particles can be directly determined from laboratory tests in the future.

As an expedient measure, primitive trial-and-error approaches can be used to solve any inverse problems. For the above-mentioned inverse problem, input parameters are the macroscopic mechanical properties of materials, while the particle-scale mechanical properties of materials, such as the particle stiffness and bond strength, are unknown variables and therefore, need to be determined. Due to the difficulty in directly solving this inverse problem, it is solved indirectly using a trial-and-error approach, in which a set of particle-scale mechanical properties of materials are assumed so that the resulting macroscopic mechanical properties can be determined from the mechanical response of the particle model having this particular set of particle-scale mechanical properties. If the resulting macroscopic mechanical properties are different from what we expected, then another set of particle-scale mechanical properties of materials are used in the particle model. This trial-and-error process needs to be continued until a set of particle-scale mechanical properties of materials can produce the expected macroscopic mechanical properties. In geological practice, a kilometer-length-scale specimen is often used to conduct a biaxial compression test and to measure the related macroscopic mechanical properties, such as the elastic modulus and material strength, from the mechanical response of the particle model with an assumed set of particle-scale mechanical properties of rocks. However, if some mechanical properties are independent of particle size or other size-dependent mechanical properties can be determined from an appropriate upscale rule, then the expected particle-scale mechanical properties of materials to be used in a particle model can be determined without a need to conduct the aforementioned trial-and-error exercise.

8.2.2 Numerical Simulation Issue Arising from Using the Explicit Dynamic Relaxation Method to Solve a Quasi-Static Problem

For the purpose of demonstrating the numerical simulation issue resulting from using the explicit dynamic relaxation method to solve a quasi-static problem, it is helpful to explain briefly how the finite element method is used to solve the same kind of problem. For the sake of simplicity, a quasi-static elastic problem is used to demonstrate the issue. Since the finite element method is based on continuum mechanics, the governing equations of a two-dimensional quasi-static elastic problem in an isotropic and homogeneous material can be expressed as follows:

$$\frac{\partial \sigma_x}{\partial x} + \frac{\partial \tau_{yx}}{\partial y} = f_x, \tag{8.20}$$

$$\frac{\partial \tau_{xy}}{\partial x} + \frac{\partial \sigma_y}{\partial y} = f_y, \tag{8.21}$$

$$\sigma_x = \frac{E(1-\nu)}{(1-2\nu)(1+\nu)} \left(\varepsilon_x + \frac{\nu}{1-\nu} \varepsilon_y \right), \tag{8.22}$$

$$\sigma_y = \frac{E(1-\nu)}{(1-2\nu)(1+\nu)} \left(\frac{\nu}{1-\nu} \varepsilon_x + \varepsilon_y \right), \tag{8.23}$$

$$\tau_{xy} = \tau_{yx} = 2G\gamma_{xy}, \tag{8.24}$$

8.2 Some Numerical Simulation Issues Associated with the Particle Simulation Method 187

$$\varepsilon_x = \frac{\partial u_s}{\partial x}, \qquad \varepsilon_y = \frac{\partial v_s}{\partial y}, \qquad \gamma_{xy} = \frac{1}{2}\left(\frac{\partial u_s}{\partial y} + \frac{\partial v_s}{\partial x}\right), \qquad (8.25)$$

where σ_x and σ_y are normal stresses of the solid matrix in the x and y directions; ε_x and ε_y are the normal strains of the solid matrix in relation to σ_x and σ_y; τ_{xy} and γ_{xy} are the shear stress and shear strain of the solid matrix; u_s and v_s are the horizontal and vertical displacements of the solid matrix; E and G are the elastic and shear modulus respectively; v is the Poisson ratio of the solid matrix; f_x and f_y are the body forces in the x and y directions;.

Note that Eqs. (8.20) and (8.21) represent the equilibrium equations, whereas Eqs. (8.22), (8.23), (8.24) and (8.25) are the constitutive equations and strain-displacement relationship equations, respectively.

By using either the variational principle or the Galerkin method (Zienkiewicz 1977, Lewis and Schrefler 1998), the above-mentioned equations can be expressed in the finite element form:

$$[K]^e \{u\}^e = \{F\}^e, \qquad (8.26)$$

where $[K]^e$ is the stiffness matrix of an element; $\{u\}^e$ and $\{F\}^e$ are the displacement and force vectors of the element.

In the finite element method, the quasi-static equilibrium problem is solved in a global (i.e. system) manner. This means that the matrices and vectors of all the elements in the system need to be assembled together to result in the following global equation:

$$[K]\{u\} = \{F\}, \qquad (8.27)$$

where $[K]$ is the global stiffness matrix of the system; $\{u\}$ and $\{F\}$ are the global displacement and force vectors of the system.

Similarly, in the distinct element method, the quasi-static equilibrium equation of a particle is of the following form:

$$[K]^p \{u\}^p = \{F\}^p, \qquad (8.28)$$

where $[K]^p$ is the stiffness matrix of a particle; $\{u\}^p$ and $\{F\}^p$ are the relative displacement and force vectors of the particle.

In order to reduce significantly the requirement for computer storage and memory, the distinct element method solves the quasi-static equilibrium equation at the particle level, rather than at the system level. This requires that a quasi-static problem be turned into a fictitious dynamic problem so that the explicit dynamic relaxation method can be used to obtain the quasi-static solution from solving the following fictitious dynamic equation:

$$[M]^p \{\ddot{u}\}^p + [K]^p \{u\}^p = \{F\}^p, \qquad (8.29)$$

where $[M]^p$ is the lumped mass matrix of the particle and $\{\ddot{u}\}^p$ is the acceleration vector of the particle.

A major difference between solving the global quasi-static equilibrium problem (i.e. Eq. (4)) in the finite element analysis and solving the fictitious dynamic problem (i.e. Eq. (8.29)) in the particle simulation is that the numerical solution to Eq. (8.26) is unconditionally stable so that it can be solved using the Gaussian elimination method and the like, while the numerical solution to Eq. (8.29) is conditionally stable if an explicit solver is used. For this reason, the critical time step, which is required to result in a stable solution for the fictitious dynamic problem, can be expressed as follows (Itasca Consulting Group, inc. 1999):

$$\Delta t_{critical} = \sqrt{\frac{m}{k}}, \tag{8.30}$$

where m is the mass of a particle and k is the stiffness between two particles.

It is immediately noted that since the value of the mass of a particle is usually much smaller than that of the stiffness of the particle, the critical time step determined from Eq. (8.30) is considerably smaller than one. This indicates that for a slow geological process of more than a few years, it may take too long to obtain a particle simulation solution. To overcome this difficulty, the scaled mass is often used in the distinct element method (Itasca Consulting Group, inc. 1999) so that the critical time step can be increased to unity or any large number if needed. However, since the scaled mass, namely the fictitious mass, is used, the time used in the distinct element method is fictitious, rather than physical. In order to remove possible chaotic behavior that may be caused by the use of arbitrarily-scaled masses, fictitious damping is also added to the particles used in a distinct element simulation (Itasca Consulting Group, inc. 1999). Thus, the numerical simulation issue resulting from using the explicit dynamic relaxation method to solve a quasi-static problem is that the time used in the distinct element method is fictitious, rather than physical.

If the mechanical response of a quasi-static system is elastic, then the solution to the corresponding quasi-static problem is unique and independent of the deformation path of the system. In this case, the explicit dynamic relaxation method is valid so that the elastic equilibrium solution can be obtained from the particle simulation using the distinct element method. However, if any failure takes place in a quasi-static system, then the quasi-static system behaves nonlinearly so that the solution to the post-failure quasi-static problem is not unique and therefore, becomes dependent on the deformation path of the system. Due to the use of both the fictitious time and the fictitious scaled mass, the physical deformation path of a system cannot be simulated using the explicit dynamic relaxation method. This implies that the post-failure particle simulation result obtained from using the explicit dynamic relaxation method may be problematic, at least from the rigorously scientific point of view. Nevertheless, if one is interested in the phenomenological simulation of the post-failure behavior of a quasi-static system, a combination of the distinct element method and the explicit dynamic relaxation method may be used to produce some useful simulation results in the engineering and geology fields (Cundall and Strack

1979, Bardet and Proubet 1992, Saltzer and Pollard 1992, Antonellini and Pollard 1995, Donze et al. 1996, Scott 1996, Strayer and Huddleston 1997, Camborde et al. 2000, Iwashita and Oda 2000, Burbidge and Braun 2002, Strayer and Suppe 2002, Finch et al. 2003, 2004, Imber et al. 2004). In this case, it is strongly recommended that a particle-size sensitivity analysis of at least two different models, which have the same geometry but different smallest particle sizes, be carried out to confirm the particle simulation result of a large-scale quasi-static geological system.

8.2.3 Numerical Simulation Issue Stemming from the Loading Procedure Used in the Particle Simulation Method

The distinct element method is based on the idea that the time step used in the simulation is chosen so small that force, displacement, velocity and acceleration cannot propagate from any particle farther than its immediate neighbors during a single time step (Potyondy and Cundall 2004). The servo-control technique (Itasca Consulting Group, inc. 1999) is often used to apply the equivalent velocity or displacement to the loading boundary of the particle model. This will pose an important scientific question: Is the mechanical response of a particle model dependent on the loading procedure that is used to apply "loads" at the loading boundary of the particle model? If the mechanical response of a particle model is independent of the loading procedure, then this issue can be neglected when we extend the application range of the particle simulation method from a small-scale laboratory test to a large-scale geological problem.

In order to answer this scientific question associated with the distinct element method that is used in the PFC2D (i.e. Particle Flow Code in Two Dimensions), it is necessary to investigate how a "load" is propagated within a particle system. For the purpose of illustrating the "load" propagation mechanism, a one-dimensional idealized model of ten particles of the same mass is considered in Fig. 8.5. The "load" can be either a directly-applied force or an indirectly-applied force due to a constant velocity in this idealized conceptual model. The question that needs to be highlighted here is that when a "load" is applied to a particle system, what is the appropriate time to record the correct response of the whole system due to this "load"? This issue is important due to the fact that particle-scale material properties of a particle are employed in a particle model and that a biaxial compression test is often used to determine the macroscopic material properties of the particle model. If the normal stiffness coefficient between any two particles has the same value (i.e. $k_i = k$ ($i = 1, 2, 3, \ldots, 10$)) and the time step is equal to the critical time step (Itasca Consulting Group, inc. 1999) in the particle model, then the displacement of particle 1 (i.e. the particle with "load" P) is P/k at the end of the first time step (i.e. $t = 1\Delta t$, where Δt is the time step). The reason for this is that in the distinct element method, the "load" can only propagate from a particle to its immediate neighboring particles within a time step. Thus, during the first time step, the other nine particles in the right part of the model are still kept at rest. This is equivalent to applying a

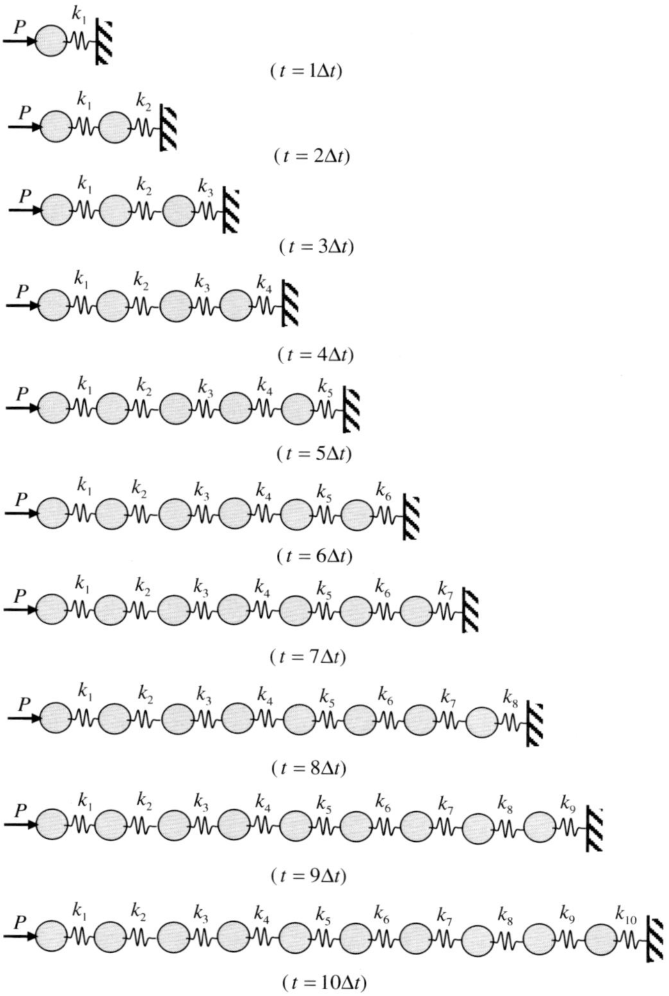

Fig. 8.5 Force propagation in a ten-particle system

fixed boundary condition at the right end of spring 1, as shown in Fig. 8.5. Similar considerations can be made for the consecutive time steps (see Fig. 8.5). Ideally, the "load" propagates through the whole system at the end of $t = 10\Delta t$, resulting in a displacement of $10P/k$ for particle 1. In general, if this one-dimensional idealized model is comprised of n particles of equal normal stiffness and mass, then the displacement of particle 1 (i.e. the particle with "load" P) is nP/k at the end of $t = n\Delta t$. Clearly, if one uses the record of the "load" and displacement at the end of the immediate loading step to determine the elastic modulus of this one-dimensional idealized particle system, then the determined elastic modulus will be exaggerated

8.2 Some Numerical Simulation Issues Associated with the Particle Simulation Method 191

by n times. Although the one-dimensional idealized particle system is a highly simplified representation of particle models, it illuminates the basic force propagation mechanism, which is valid for two- and three-dimensional particle models.

The conventional loading procedure used in the distinct element method is shown in Fig. 8.6. In order to reduce inertial forces exerted on the loading-boundary particles due to a suddenly-applied velocity at the first loading step, an improved-conventional loading procedure is also used in the distinct element method (see Fig. 8.6). Since both the conventional loading procedure and the improved-conventional loading procedure are continuous loading procedures, it is impossible to take the correct record of the "displacement", just at the end of a "load" increment. In other words, when a "load" increment is applied to the particle system, it takes a large number of time steps for the system to reach a quasi-static equilibrium state. It is the displacement associated with the quasi-static equilibrium state that represents the correct displacement of the system due to this particular "load" increment. For this reason, a new discontinuous loading procedure is proposed in this section. As shown in Fig. 8.6, the proposed loading procedure comprises two main types of periods, a loading period and a frozen period. Note that the proposed loading procedure shown in this figure is illustrative. In real numerical practice, a loading period is only comprised of a few time steps to avoid the

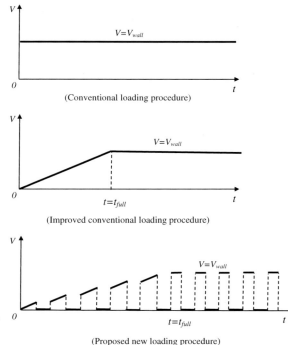

Fig. 8.6 Illustration of different loading procedures for the loading of a particle model

occurrence of any unphysical damage/crack in the particle model, whereas a frozen period can be comprised of thousands of time steps. In the loading period, a velocity increment is applied to the loading boundary of the system, while in the frozen period the loading boundary is fixed to allow the system to reach a quasi-static equilibrium state after a large number of time steps. It needs to be pointed out that a pair of load and displacement (or stress and strain) is correctly recorded at the end of a frozen period.

Based on the above-mentioned conceptual understanding, the theoretical expression of the proposed loading algorithm can be deduced as follows. An applied force in the quasi-static system can be divided into M loading increments.

$$P = \sum_{i=1}^{M} \Delta P_i, \qquad (8.31)$$

where P is the applied force; ΔP_i is the *ith* loading force increment and M is the total number of loading steps.

For each loading force increment, ΔP_i, it is possible to find a solution, ΔS_i, which satisfies the following condition:

$$\lim_{j \to N_i} (\Delta S_{ij} - \Delta S_{ij-1}) = 0, \qquad (8.32)$$

where ΔS_{ij} is the solution at the *jth* time step due to the *ith* loading force increment; ΔS_{ij-1} is the solution at the *(j-1)th* time step due to the *ith* loading force increment; N_i is the total number of time steps for the particle system to reach a quasi-static equilibrium state after the application of the *ith* loading force increment to the system.

From the numerical analysis point of view, Equation (8.32) can be straightforwardly replaced by the following equation:

$$\max \left| \frac{\Delta S_{ij} - \Delta S_{ij-1}}{\Delta S_{ij}} \right| < \delta \qquad (\text{at } j = N_i), \qquad (8.33)$$

where δ is the tolerance of the solution accuracy.

It needs to be pointed out that the value of N_i can be determined using the convergent condition expressed by Eq. (8.33), so that different values of N_i may be obtained in the numerical simulation. It is the convergent condition that approximately warrants the solution of the quasi-static nature, if the value of δ is not strictly equal to zero.

Thus, the total solution S corresponding to the applied force P can be expressed as

$$S = \sum_{i=1}^{M} \Delta S_i. \qquad (8.34)$$

8.2 Some Numerical Simulation Issues Associated with the Particle Simulation Method

It is noted that the total number of time steps to produce the total solution S is calculated using the following formula:

$$N_{total} = \sum_{i=1}^{M} N_i. \tag{8.35}$$

Similarly, at the end of each loading step, there exist the following relationships:

$$P_j = \sum_{i=1}^{j} \Delta P_i \quad (j = 1, 2, \ldots, M), \tag{8.36}$$

$$S_j = \sum_{i=1}^{j} \Delta S_i \quad (j = 1, 2, \ldots, M), \tag{8.37}$$

where P_j is the total applied loading force at the end of loading step j; S_j is the total solution at the end of loading step j. This indicates that at the end of a loading step, a point (P_j, S_j) has been obtained in the load-solution space. Therefore, at the end of loading step M, we have obtained M points so that it is possible to link all these M points together to obtain a load-solution path in the load-solution space. Clearly, if P and S represent the applied loading force and resulting displacement, then a force-displacement curve is obtained in the force-displacement space. Alternatively, if P and S stand for the applied loading stress and resulting strain respectively, then a stress-strain curve is obtained in the stress-strain space.

With the biaxial compression test of a particle simulation model taken as an example, the general steps of using the proposed loading algorithm can be summarized as follows. (1) For a given load increment, which needs to be applied to the boundary of a particle model, the servo-control technique (Itasca Consulting Group, inc. 1999) is used to apply the equivalent velocity to the boundary of the particle model. This means that an equivalent velocity is applied to the boundary of the particle model for a few time steps during the particle simulation. The number of time steps, during which the equivalent velocity needs to be applied to the boundary of the particle model, depends on the desired load increment that is applied to the boundary of the particle model. This step is called the loading step and the corresponding loading time or the number of related time steps is called the loading period. It is noted that in theory, a loading period should be as small as possible, so that any unphysical damage/crack can be prevented from occurring during the loading period. (2) After a loading period, the applied equivalent velocity is set to be zero so that the loading boundary of the particle model becomes frozen (i.e. fixed). This step is called the frozen step. During a frozen step, the particle simulation model is kept running until an equilibrium state is reached. Therefore, the duration of a frozen period depends on how quickly the particle simulation model can reach its corresponding equilibrium state. (3) When the particle simulation model reaches an equilibrium state, an applied load increment such as a stress increment on the loading boundary of the particle model can be calculated so that the value of the desired load increment is determined. At the same time, other quantities of interest such as a displacement increment, strain increment and so forth can be also determined. Thus, a point relating stress to strain in a stress-strain curve is obtained at the end of this step. For this reason, this step is called the result acquisition step. (4) Steps 1–3 are repeated for every desired load increment until the final stage of the particle simulation is reached. As a result, many points relating stress to strain in a stress-strain curve have been obtained at the end of the particle simulation. (5) Finally, all the obtained points relating stress to strain are connected to generate a stress-strain curve, which should represent the true quasi-static behavior for the particle simulation model.

8.3 An Upscale Theory of Particle Simulation for Two-Dimensional Quasi-Static Problems

For the purpose of establishing an upscale theory associated with the particle simulation of two-dimensional quasi-static systems, it is necessary to understand the particle-scale mechanical properties and their relations to macroscopic mechanical properties, which are available from either a laboratory test or an *in-situ* measurement. If circular particles of unit thickness are used in the simulation of a particle assembly, the particle-scale mechanical properties such as the stiffness and bond strength of the particle are needed for a contact-bond model. Since it is very difficult, if not impossible, to directly measure particle-scale mechanical properties from laboratory tests, it is common practice to determine these particle-scale mechanical properties from the macroscopic mechanical properties such as the elastic modulus, tensile and shear strength of particle materials. From the analog of a two-circle contact with an elastic beam (Itasca Consulting Group, inc. 1999), it has been demonstrated that there may exist an upscale rule, which states that the contact stiffness of a circular particle is only dependent on the macroscopic elastic modulus and independent of the diameter of the circular particle. The value of the contact stiffness of a circular particle is equal to twice that of the macroscopic elastic modulus of the material. On the other hand, the contact bond strength of a circular particle is directly proportional to both the tensile/shear strength of the particle material and the diameter of the circular particle.

In order to facilitate the derivation of the corresponding similarity criteria, it is assumed that the problem domain is comprised of a homogeneous medium. Since a heterogeneous medium can be divided into many sub-domains of homogeneous materials, the derived similarity criteria in this investigation is, as demonstrated later by the test and application example in this chapter, valid and applicable for any two similar particle models of heterogeneous media, as long as each homogeneous sub-domain of the two similar particle models satisfies the required geometrical similarity criterion. For this reason, the proposed upscale theory associated with the particle simulation is also applicable for the particle simulation of a geometrically-similar

8.3 An Upscale Theory of Particle Simulation for Two-Dimensional Quasi-Static Problems

geological medium, which may be non-homogeneous due to the presence of faults, cracks and large geological structures, provided that the particle model of the geological medium satisfies the required geometrical similarity criterion.

For the analog of a two-circle contact with an elastic beam (Itasca Consulting Group, inc. 1999), it is assumed that the behaviour of the contact between two particles is equivalent to that of an elastic beam with its ends at the two particle centres. The beam is loaded at its ends by the force acting at the centre of each particle. Under this assumption, the stress of the beam can be expressed as follows:

$$\sigma = \frac{P}{D} = E\varepsilon = E\frac{\Delta}{D}, \tag{8.38}$$

where σ and ε are the stress and strain of the equivalent elastic beam; P is the force acting at the centre of each particle; E is the elastic modulus of the particle material; Δ is the deformation of the equivalent elastic beam; D is the diameter of the particle.

Since the two particles are connected in series, the deformation of the equivalent elastic beam can be also expressed as follows:

$$\Delta = \frac{2P}{k_n}, \tag{8.39}$$

where k_n is the normal stiffness of a particle.

Substituting Eq. (8.39) into Eq. (8.38) yields the following equation:

$$k_n = 2E. \tag{8.40}$$

Since the contact-bond strength is expressed in the unit of force, the following relationships exist mathematically:

$$b_n = \alpha\bar{\sigma}_n D, \tag{8.41}$$

$$b_s = \alpha\bar{\tau}_s D, \tag{8.42}$$

where b_n and b_s are the normal and tangential bond strengths at a contact between two particles; D is the diameter of the particle; α is a constant; $\bar{\sigma}_n$ and $\bar{\tau}_s$ are the unit normal and tangential contact bond strengths, which are defined as the normal and tangential contact bond strengths per unit length of the particle diameter (Zhao et al. 2007b). If α is assumed to be equal to one, then the values and units of the unit normal and tangential contact bond strengths are exactly the same as those of the macroscopic tensile and shear strengths of the particle material, while the values of the normal and tangential contact bond strengths of a particle are equal to the product of the unit normal/tangential contact bond strengths and the diameter value of the particle. Thus, the variation of the normal/tangential contact bond strength of a particle with its diameter can be straightforwardly considered in the particle simulation.

Note that in the particle simulation method such as the distinct element method, the contact force exerted on a particle is calculated using the following formula:

$$F_n = k_n u_n, \qquad (8.43)$$

where F_n is the normal contact force at a contact between two particles; u_n is the normal displacement at the contact. It needs to be pointed out that in the linear elastic range of the particle material, a similar relationship to Eq. (8.43) is also valid for the shear contact force and tangential displacement at a contact between two particles.

For a quasi-static system, two particle models of different length-scales can be considered to establish the upscale theory. The first particle model (i.e. model one) is of a small length-scale such as a laboratory length-scale, while the second particle model (i.e. model two) is of a large length-scale such as a regional geological length-scale. From elasticity theory, the necessary condition, under which these two models are similar, is that the relative displacements (i.e. strain) of the two models are identical. This results in the following similarity criterion for two particle models of the same number of particles:

$$\frac{u_n^{m1}}{u_n^{m2}} = \frac{D^{m1}}{D^{m2}} = \frac{L^{m1}}{L^{m2}}, \qquad (8.44)$$

where u_n^{m1} and u_n^{m2} are the displacements of models one and two; D^{m1} and D^{m2} are the diameters of the particles of models one and two respectively; L^{m1} and L^{m2} are the geometrical lengths of the two models.

Equation (8.44) indicates that if two particle models of different length-scales are similar, then both the displacement ratio and the diameter ratio of the two models are equal to their geometrical length ratio. For this reason, Eq. (8.44) is called the first similarity criterion between two particle models of different length-scales.

Consideration of Eqs. (8.40), (8.43) and (8.44) yields the following similarity criterion for the two particle models:

$$\frac{F_n^{m1}}{F_n^{m2}} = \frac{u_n^{m1}}{u_n^{m2}} = \frac{L^{m1}}{L^{m2}}, \qquad (8.45)$$

where F_n^{m1} and F_n^{m2} are the contact forces between two particles of models one and two respectively.

Equation (8.45) states that the similarity ratio of the contact forces of the two models is equal to both their similarity ratio of displacements and their similarity ratio of geometrical lengths. This equation is called the second similarity criterion between two particle models of different length-scales.

8.3 An Upscale Theory of Particle Simulation for Two-Dimensional Quasi-Static Problems

Similarly, consideration of Eqs. (8.41) and (8.42) results in the following similarity criterion for the two particle models:

$$\frac{b_n^{m1}}{b_n^{m2}} = \frac{b_s^{m1}}{b_s^{m2}} = \frac{D^{m1}}{D^{m2}} = \frac{L^{m1}}{L^{m2}}, \tag{8.46}$$

where b_n^{m1} and b_n^{m2} are the normal bond strengths of the particles of models one and two; b_s^{m1} and b_s^{m2} are the tangential bond strengths of the particles of models one and two respectively.

Equation (8.46) indicates that the similarity ratios of the (normal and tangential) bond strengths of the two particle models are equal to both their similarity ratio of particle diameters and their similarity ratio of geometrical lengths. This equation is called the third similarity criterion between two particle models of different length-scales. Since the similarity ratios of the particle bond strengths are equal to the similarity ratio of the particle contact forces, the occurrence of the first failure should be similar for two similar particle models. This indicates that if the first failure occurs at a particle in model one, then the first failure should occur at a similar counterpart in model two.

For a quasi-static geological system, it is important to consider the gravity effect in two particle models of different length-scales. At the particle level, the gravity force exerted on a circular particle can be expressed as follows:

$$G_p = \frac{\pi}{4} \rho_p g D^2 T, \tag{8.47}$$

where G_p is the gravity force exerted on a particle; ρ_p is the density of the particle material; g is the gravity acceleration; T is the thickness of the circular particle. Note that $T \equiv 1$ in this investigation.

From Eq. (8.47), the similarity ratio of gravity forces for the two particle models of different length-scales can be derived and expressed as follows:

$$\frac{G_p^{m1}}{G_p^{m2}} = \frac{\rho_p^{m1}}{\rho_p^{m2}} \frac{g^{m1}}{g^{m2}} \frac{(D^{m1})^2}{(D^{m2})^2}, \tag{8.48}$$

where G_p^{m1} and G_p^{m2} are the gravity forces exerted on the particles of models one and two respectively; ρ_p^{m1} and ρ_p^{m2} are the densities of the particle materials of the two models; g^{m1} and g^{m2} are the gravity accelerations of the particles of models one and two respectively.

In order to implement this similarity criterion in particle simulation models, it is desirable to keep the similarity ratio of the gravity forces equal to that of the geometrical lengths of the two similar particle models. Since the explicit dynamic relaxation method is used to solve the equation of motion in a particle simulation (Itasca Consulting Group, inc. 1999), the similarity ratio of the particle densities needs to be equal to one so that it does not affect the time-step similarity of the two models, as discussed below. For this reason, it is necessary to use the following alternative relationships:

$$\frac{g^{m1}}{g^{m2}} = \frac{D^{m2}}{D^{m1}}, \quad \frac{\rho_p^{m1}}{\rho_p^{m2}} = 1, \tag{8.49}$$

$$\frac{G_p^{m1}}{G_p^{m2}} = \frac{\rho_p^{m1}}{\rho_p^{m2}} \frac{g^{m1}}{g^{m2}} \frac{(D^{m1})^2}{(D^{m2})^2} = \frac{D^{m1}}{D^{m2}}. \tag{8.50}$$

Clearly, Eq. (8.49) states that in order to maintain the similarity of two particle models of different length-scales, the similarity ratio of the gravity accelerations should be equal to the inverse of the similarity ratio of the geometrical lengths for the two similar particle models. In this regard, Eq. (8.49) is called the fourth similarity criterion between two particle models of different length-scales.

For the particle simulation based on the distinct element method, the numerical solution to the equation of motion is conditionally stable because the explicit dynamic relaxation solver is used. The critical time-step, which is required to result in a stable solution, can be expressed as follows (Itasca Consulting Group, inc. 1999):

$$\Delta t_{critical} = \sqrt{\frac{m}{k}}. \tag{8.51}$$

where $\Delta t_{critical}$ is the critical time-step; m is the mass of a particle and k is the stiffness between two particles.

It is immediately noted from Eq. (8.51) that the critical time-steps, which are used in two similar particle models of different length-scales, satisfy the following similarity criterion:

$$\frac{\Delta t_{critical}^{m1}}{\Delta t_{critical}^{m2}} = \sqrt{\frac{\rho_p^{m1}}{\rho_p^{m2}} \frac{D^{m1}}{D^{m2}}} = \frac{D^{m1}}{D^{m2}}, \tag{8.52}$$

where $\Delta t_{critical}^{m1}$ and $\Delta t_{critical}^{m2}$ are the corresponding critical time-steps used in models one and two respectively.

Equation (8.52) provides an auxiliary similarity criterion, which is a direct result from the above-mentioned fourth similarity criterion, for the two similar particle models of different length-scales.

It needs to be pointed out that the proposed upscale theory associated with particle simulation methods is strictly valid when the mechanical response of a two-dimensional particle assembly is within the elastic range. If the loading increment is small enough to prevent any unphysical damage/crack from occurring within a two-dimensional particle assembly and the number of time-steps is large enough to enable the particle assembly to reach a quasi-static state during this loading period, which can be achieved using the newly-proposed loading procedure associated with the distinct element method (Zhao et al. 2007b), the proposed upscale theory for the particle simulation is also approximately valid for the post-failure response of the particle assembly, as demonstrated in the next section.

8.4 Test and Application Examples of the Particle Simulation Method

As shown in Fig. 8.7, two samples of different sizes are considered in the particle simulation tests. The first test sample is of a small size (1 by 2 m) and is simulated using 1000 particles. It was noted that if a regular hexagonal lattice, in which a particle is in contact with its six neighboring particles, is used in the particle simulation (Donze et al. 1994), an unphysical, first-order geometrical control may result in well-defined 60° planes of weakness in the lattice. These planes can further control the geometry of the resulting failure (Donze et al. 1994). To prevent the above-mentioned effect from taking place, we generate a particle model in which the particles are distributed randomly so that the likelihood for the occurrence of the preferred planes of weakness can be eliminated. The maximum and minimum radii of particles are approximately 0.0172 m and 0.0115 m, resulting in an average radius of 0.0144 m. On the other hand, the second test sample is of large size (1 by 2 km) and is also simulated using 1000 particles. The maximum and minimum radii of particles are approximately 17.24 m and 11.49 m, resulting in an average radius of 14.37 m. The initial porosity of both the small and the large test samples is set to be 0.17 in the particle simulation. The density of the particle material is 2500 kg/m^3 and the friction coefficient of the particle material is 0.5, while the confining stress is assumed to be 10 MPa in the following numerical experiments. Due to a significant

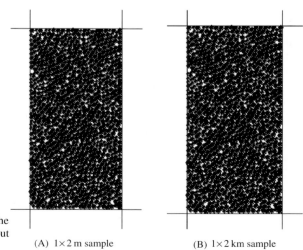

Fig. 8.7 Two samples of the same number of particles but different sizes (A) 1×2 m sample (B) 1×2 km sample

size difference between the small and large test samples, the size effect of the test sample can be investigated through the particle simulation.

The stiffness and bond strength of particles in a test sample can be predicted using the macroscopic mechanical properties such as the elastic modulus, tensile and shear strength of particle materials. From the analog of a two-circle contact with an elastic beam (Itasca Consulting Group, inc. 1999), it has been demonstrated that there may exist an upscale rule, which states that the contact stiffness of a circular particle is only dependent on the macroscopic elastic modulus and independent of the diameter of the circular particle. The value of the contact stiffness of a circular particle is equal to twice that of the macroscopic elastic modulus of the material. On the other hand, the contact bond strength of a circular particle is directly proportional to both the tensile/shear strength of the particle material and the diameter of the circular particle. To reflect this fact, the concept of the unit normal/tangential contact bond strength, which is defined as normal/tangential contact bond strength per unit length of the particle diameter, is used in this study. Using this definition, the value of the unit normal/tangential contact bond strength is equal to that of the macroscopic tensile/shear strength of the particle material, while the value of the normal/tangential contact bond strength of a particle is equal to the product of the unit normal/tangential contact bond strength and the diameter of the particle. In this way, the variation of the normal/tangential contact bond strength of a particle with its diameter is considered in the particle simulation. Keeping the above considerations in mind, the following macroscopic mechanical properties of rock masses are used to determine the contact mechanical properties of the particle material used in the simulation. The macroscopic elastic modulus of the particle material is 0.5 GPa, resulting in a contact stiffness (in both the normal and the tangential directions) of 1.0 GN/m for both the small and the large test samples. The macroscopic tensile strength of the particle material is 10 MPa, while the macroscopic shear strength of the particle material is 100 MPa for both the small and the large test samples. The loading period is 10 time-steps, while the frozen period is 9990 time-steps in the numerical biaxial compression tests.

For the particle simulation associated with the distinct element method (Itasca Consulting Group, inc. 1999), crack initiation and generation are determined using the following crack criteria:

$$F_{Cn} \geq b_n, \qquad (8.53)$$

$$|F_{Cs}| \geq b_s, \qquad (8.54)$$

where F_{Cn} and F_{Cs} are the normal and shear contact forces at a contact between any two particles; b_n and b_s are the normal and tangential bond strengths at the contact between the two particles.

Obviously, Eq. (8.53) indicates that if a normal contact force exceeds the corresponding normal tensile bond strength at a contact between two particles, the normal tensile bond is broken and therefore, a tensile crack is generated at the contact.

8.4 Test and Application Examples of the Particle Simulation Method   201

Similarly, Eq. (8.54) indicates that if the absolute value of a shear contact force exceeds the corresponding tangential bond strength at a contact between two particles, the tangential shear bond is broken and therefore, a shear crack is generated at the contact. The above-mentioned crack criteria are checked for each time step during the particle simulation of a computational model.

8.4.1 Comparison of the Proposed Loading Procedure with the Conventional Loading Procedure

Figure 8.8 shows the effect of the loading rate and sample size on the deviatoric stress versus axial strain curve for both the small and the large samples of 1000 particles. The deviatoric stress is defined as the axial stress minus the confining stress in this investigation. As usual, the servo-control technique (Itasca Consulting Group, inc. 1999) is used to apply the equivalent velocity to the loading boundary of the particle model. The equivalent velocity is called the loading rate hereafter. It is obvious that the simulated stress-strain curve is independent of the two loading rates (i.e. LR = 1.0 m/s and LR = 10 m/s in this figure) and sample sizes in the elastic response range, where there is no occurrence of any failure in the test material. It can be found, from the stress-strain curve, that the simulated elastic modulus of the particle material is equal to 0.5 GPa, which is identical to the desired value of the expected macroscopic elastic modulus of the particle material. This indicates that the upscale rule established from the analog of a two-circle contact with an elastic beam (Itasca Consulting Group, inc. 1999) is appropriate for predicting the elastic modulus when the proposed loading procedure is used in the simulation of a two-dimensional particle model.

It is worth pointing out that since the equivalent velocity is simultaneously applied to both the upper and the lower boundaries of a test sample, the strain rate of the sample is equal to the ratio of the loading rate to the half-length of the sample. In the case of the small sample of 1000 particles, the corresponding strain rates of the sample are 10 (1/s) and 1.0 (1/s) for loading rates of 10 (m/s) and 1.0 (m/s) respectively. However, in the case of the large sample of 1000 particles, the corresponding strain rates of the sample are 0.01 (1/s) and 0.001 (1/s) for the same loading rates (i.e. 10 (m/s) and 1.0 (m/s)) as those used in the small sample. It can be observed that due to the solution uniqueness of the samples in the elastic range, all the stress-strain curves of both the small and the large samples are identical before the first failure takes place in these two samples. Although the post-failure stress-strain curves of the two samples are very similar in shape, they are not identical for both the small and the large samples, indicating that the post-failure response of a sample is also dependent on the strain rate of the sample. This issue needs to be considered when a simulation result is obtained from a particle model.

Next, we compare the particle simulation results obtained from using the proposed loading procedure with those obtained from using the improved conventional loading procedure. For this purpose, both the small and the large test samples of

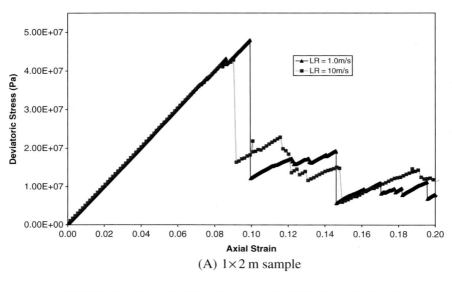

(A) 1×2 m sample

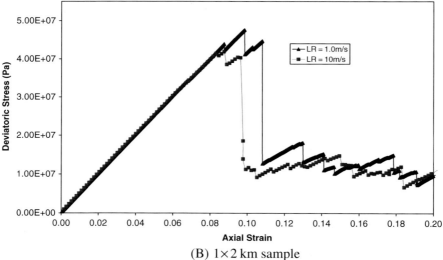

(B) 1×2 km sample

Fig. 8.8 Effects of loading rate and sample size on the deviatoric stress versus axial strain curve using the proposed loading procedure

1000 particles are used to conduct biaxial compression tests using the improved conventional loading procedure. Figure 8.9 shows the effect of the loading rate on the curve of deviatoric stress versus axial strain for both the small and the large test samples of 1000 particles using the improved conventional loading procedure. As mentioned previously, the strain rates of both the small and the large test samples are

8.4 Test and Application Examples of the Particle Simulation Method 203

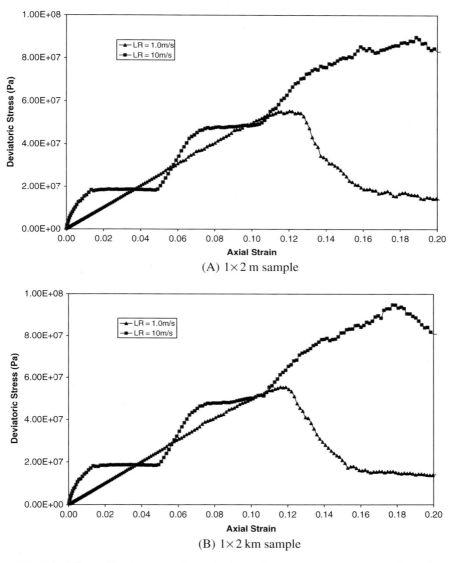

Fig. 8.9 Effects of loading rate and sample size on the deviatoric stress versus axial strain curve using the improved conventional loading procedure

different, even though the loading rates of these two samples are identical. It is obvious that the general solution pattern for both the small and the large test samples of 1000 particles is, even though not identical, very similar. This indicates that the sample size of a particle model has little influence on the mechanical response of the model in the elastic range of the particle material, even if the improved conventional loading procedure is used to produce the simulation results. However, the mechanical responses of both the small and the large test samples of 1000 particles are clearly dependent on the loading rate in the elastic range of the particle material. In the case of a loading rate of 10 m/s, there is an oscillatory behaviour in the stress-strain curve. Such an oscillatory behaviour does not occur in Fig. 8.8, where the mechanical responses of both the small and the large test samples of 1000 particles are obtained from using the proposed loading procedure. Nevertheless, the oscillatory behaviour of the mechanical response obtained from using the improved conventional loading procedure is greatly reduced when the smaller loading rate (i.e. LR = 1.0 m/s) is used in the particle simulation, indicating that the use of the improved conventional loading procedure in a particle simulation may produce some useful results as long as the loading rate is kept very small in the simulation. In the case of a loading rate of 10 m/s, the maximum yielding strength obtained from using the improved conventional loading procedure is almost twice that obtained from using the proposed loading procedure, implying that the maximum yielding strength can be overestimated when using the improved conventional loading procedure. It is interesting to note that the mechanical response obtained from using the improved conventional loading procedure exhibits stronger ductile behavior (Fig. 8.9), while the mechanical response obtained from using the proposed loading procedure exhibits stronger brittle behavior (Fig. 8.8) for exactly the same test sample. This demonstrates that in addition to conceptual soundness, the proposed loading procedure is more appropriate than the improved conventional loading procedure in dealing with the numerical simulation of the brittle behavior of crustal rocks.

8.4.2 The Similarity Test of Two Particle Samples of Different Length-Scales

The same two samples of different length-scales are considered here. The first test sample is of small size (1 by 2 m) and is simulated using 1000 randomly-distributed particles. The maximum and minimum radii of the particles used in the particle sample are approximately 0.01724 m and 0.01149 m, resulting in an average radius of 0.01437 m. On the other hand, the second test sample is of large size (1 by 2 km) and is also simulated using 1000 randomly-distributed particles. The second test sample is artificially designed to validate the proposed upscale theory in this study. The maximum and minimum radii of the particles used in the particle sample are approximately 17.24 m and 11.49 m, resulting in an average radius of 14.37 m. Since the similarity ratio (i.e. 1/1000) of particle diameters is equal to that of geometrical lengths for the two samples, the first similarity criterion is satisfied between these two test samples. The initial porosity of both the small and the large test samples is set to be 0.17 in the particle simulation. The density of the particle material is 2500 kg/m^3 and the friction coefficient of the particle material is 0.5, while the confining stress is taken as 10 MPa in the following numerical experiments. The macroscopic elastic modulus of the particle material is 0.5 GPa, resulting in a

8.4 Test and Application Examples of the Particle Simulation Method

contact stiffness (in both the normal and the tangential directions) of 1.0 GN/m for each particle in both the meter-scale and the kilometer-scale test samples. Since the same value of the contact stiffness is used in both the meter-scale and kilometer-scale samples, the second similarity criterion is satisfied between these two test samples.

To satisfy the third similarity criterion, it is assumed that the macroscopic tensile strength of the particle material is 10 MPa, while the macroscopic shear strength of the particle material is 100 MPa for both the meter-scale and the kilometer-scale test samples. In the case of $\alpha = 1$, the values of the unit normal and tangential contact bond strengths are equal to the macroscopic tensile and shear strengths of the particle material (Zhao et al. 2007b). Since the normal and tangential contact bond strengths are directly proportional to the particle diameter, the third similarity criterion is satisfied between these two test samples. Thus, all the necessary similarity criteria are satisfied for these two samples of different length-scales. For the purpose of testing the upscale theory in a wide parameter space, effects of three important parameters, such as the confining stress, the normal bond strength and the shear bond strength, are investigated using the two similar samples of different length-scales.

Figure 8.10 shows the effect of the confining stress on the deviatoric stress versus axial strain curve for both the meter-scale and the kilometer-scale samples of 1000 particles. Keeping other parameters unchanged, three different values of the confining stress, namely CS $= 0.1$, 1 and 10 MPa shown in Fig. 8.10, are considered in the particle simulation of the two similar test samples. Note that the deviatoric stress is defined as axial stress minus confining stress in this investigation. Because it is difficult to directly apply a stress boundary condition to the boundary of a particle model, the servo-control technique (Itasca Consulting Group, inc. 1999) is used, as an alternative, to apply the equivalent velocity of the applied stress to the loading boundary of the particle model. Using the newly-proposed loading procedure associated with the distinct element method (Zhao et al. 2007b), the equivalent velocity of 1 m/s is applied to both the upper and the lower boundaries of the two similar test samples. It is obvious that the simulated stress-strain curve is dependent on the confining stress. In particular, the maximum values of the failure stresses of the particle samples are significantly different for three different confining stresses. The general trend of the confining stress effect is that the higher the confining stress, the greater the maximum value of the failure stress of the particle sample. Since both the meter-scale sample and the kilometer-scale sample are similar, the simulated stress-strain curves are similar for these two samples of different length-scales, especially in the elastic response range of the particle assemblies. This demonstrates that the proposed upscale theory is appropriate and useful for establishing an intrinsic relationship between two similar particle systems of different length-scales.

To examine whether or not the similar particle samples can reproduce the rock dilation phenomenon observed from laboratory experiments, the effect of the confining stress on the dilation of a particle model is also considered in the particle simulation of the test sample. Figure 8.11 shows the effect of the confining stress on the volumetric strain versus axial strain curve for both the meter-scale

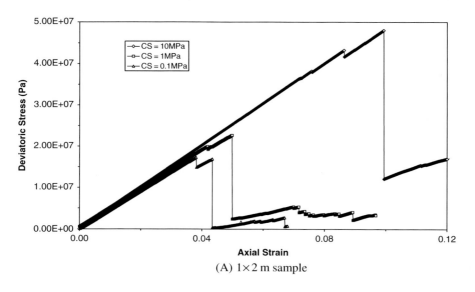

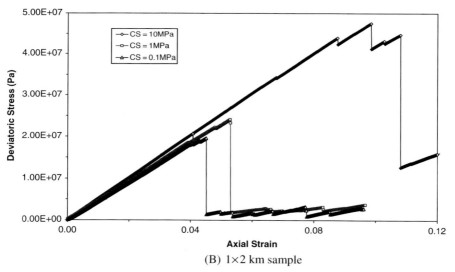

Fig. 8.10 Effects of confining stress on the curve of deviatoric stress versus axial strain

and the kilometer-scale samples. In this figure, the compressive axial strain is positive, while the expansive volumetric strain is considered to be positive. This means that a positive dilation stands for a volumetric expansion, whereas a negative dilation represents a volumetric compression of a particle sample. It can be observed that the confining stress has a significant influence on the dilation of both the meter-scale and the kilometer-scale particle samples. At the early stage of the particle

8.4 Test and Application Examples of the Particle Simulation Method

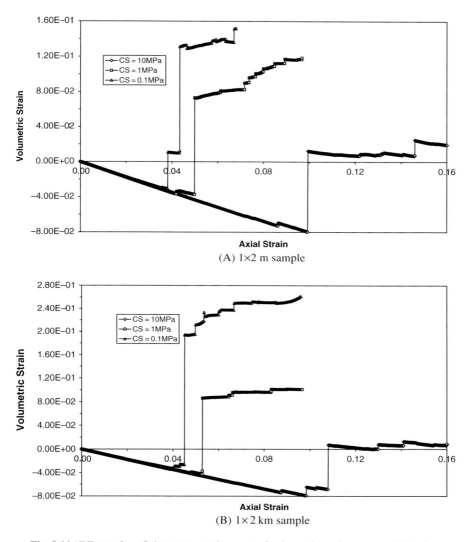

Fig. 8.11 Effects of confining stress on the curve of volumetric strain versus axial strain

simulation, the dilation of the particle sample is negative, implying that the total volume of the particle sample decreases with an increase of the axial strain. This phenomenon continues until the mechanical response of the particle sample reaches a critical stage, at which a major failure takes place within the particle sample so that there is a remarkable increase in the volumetric strain of the particle sample. After this major failure, the volumetric strain of the particle sample becomes positive, implying that the volume expansion takes place within the particle sample. The dilation phenomenon of the particle sample is consistent with what was observed in the laboratory experiments of rocks (Jaeger and Cook 1976). Since higher confining stress can effectively prevent the lateral expansive axial stain of a test sample from occurring at the critical stage, the value of the volumetric strain in the case of the confining stress being 10 MPa is almost double the value of the volumetric strain in the case of the confining stress being 0.1 MPa. This indicates that an increase in the confining stress can result in an increase in the compressive volumetric strain at the critical stage of the test sample. Clearly, the curves of volumetric strain versus axial strain (as shown in Fig. 8.11) are identical in the elastic response ranges of both the meter-scale and the kilometer-scale samples, while they are very similar in the post-failure response ranges of the two samples of different length-scales. It is noted that in the case of the confining stress being 0.1 MPa, there is a considerable discrepancy between the maximum values of the volumetric strain in the post-failure response ranges of the two samples of different length-scales. Nevertheless, this discrepancy is significantly reduced in the case of the confining stress being increased to 10 MPa. This further demonstrates that the proposed upscale theory is appropriate and useful for establishing an intrinsic relationship between two similar particle systems of different length-scales.

Next, we investigate the effect of the normal bond strength of particles on the mechanical responses of both the meter-scale and the kilometer-scale samples. In this case, the confining stress is taken as 10 MPa, while the unit shear bond strength of particles is 100 MPa for both the test samples. Three different values of the unit normal bond strengths of particles, namely NB = 0.1 MPa, 1 MPa and 10 MPa (as shown in Figs. 8.12 and 8.13), are used in the particle simulation of the two similar test samples.

Figure 8.12 shows the effect of the normal bond strength on the curves of deviatoric stress versus axial strain, while Fig. 8.13 shows the effect of the normal bond strength on the curves of volumetric strain (i.e. the dilation) versus axial strain for both the meter-scale and the kilometer-scale samples. Due to the geometrical similarity between these two samples, the simulation results from the meter-scale sample are very similar to those from the kilometer-scale sample, especially in the elastic response ranges of the two similar samples. It is also noted that the normal bond strength of particles has a significant effect on both the stress-strain and the dilation-strain curves of the two similar samples. The general trend is that the maximum failure stress of a particle sample increases with an increase in the normal bond strength of the particles used in the particle sample.

Similarly, the effect of the shear bond strength on the mechanical response of the two similar samples is examined by considering three different values of the unit shear bond strengths, namely SB = 1, 10 and 100 MPa (as shown in Figs. 8.14 and 8.15). In this situation, both the confining stress and the unit normal bond strength are kept as two different constants, which are equal to 10 and 1 MPa in the particle simulation. Figures 8.14 and 8.15 show the effect of the shear bond strength of particles on the stress-strain and dilation-strain curves for both the meter-scale and the kilometer-scale samples respectively. In addition to a clear similarity between the simulation results obtained from the two similar particle samples of different length-scales, it is interesting to note that in the case of SB = 100 MPa, both the

8.4 Test and Application Examples of the Particle Simulation Method

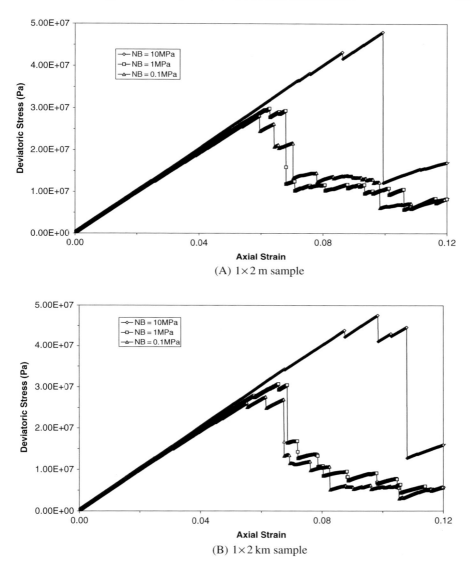

Fig. 8.12 Effects of normal bond strength on the curve of deviatoric stress versus axial strain

two samples exhibit a strong brittle behaviour, while in the case of SB = 1 MPa, the same two samples exhibit a strong ductile behaviour, indicating that the shear bond strength of particles has a significant effect on the brittle and ductile behaviour of a particle assembly. This recognition can provide a basic guidance for the selection of the shear bond strength of the particle material.

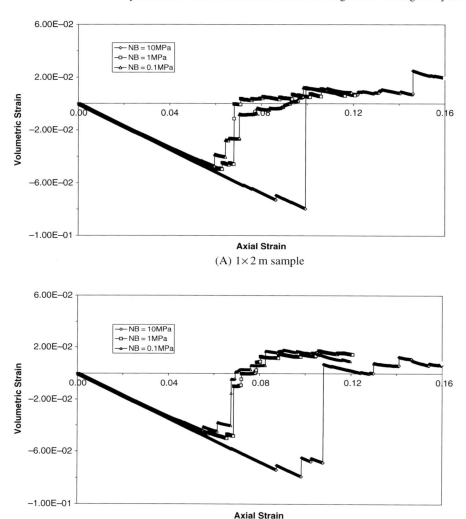

Fig. 8.13 Effects of normal bond strength on the curve of volumetric strain versus axial strain

8.4.3 Particle Simulation of the Folding Process Using Two Similar Particle Models of Different Length-Scales

As the further test and application example, two particle models of different length-scales, as shown in Fig. 8.16, are used to simulate the folding process that occurs within the upper crust of the Earth. In the first model, the length and thickness of the crust are 20 m and 2.5 m, while in the second model, the length and thickness

8.4 Test and Application Examples of the Particle Simulation Method 211

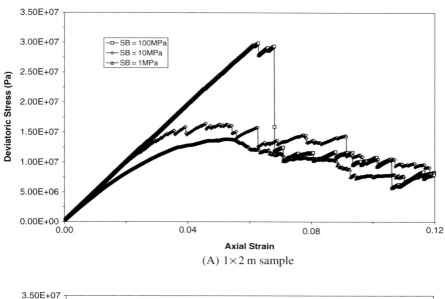

(A) 1×2 m sample

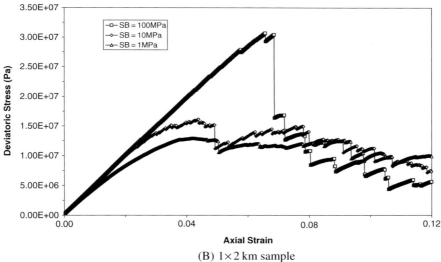

(B) 1×2 km sample

Fig. 8.14 Effects of shear bond strength on the curve of deviatoric stress versus axial strain

of the crust are 20 km and 2.5 km respectively. The first particle model is artificially designed to validate the proposed upscale theory in this investigation. Each particle model has 8000 randomly-distributed particles. The minimum and maximum radii of the particles used in the meter-scale model are 0.020318 m and 0.030477 m, resulting in an average radius of 0.025397 m. On the other hand, the minimum and maximum radii of the particles used in the kilometer-scale model are 20.318 m

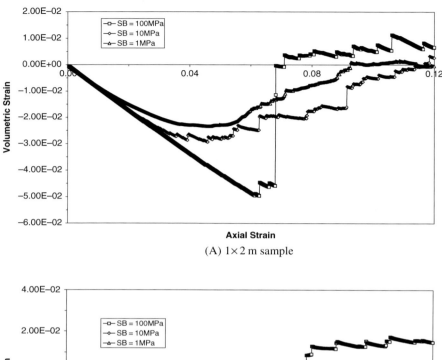

Fig. 8.15 Effects of shear bond strength on the curve of volumetric strain versus axial strain

and 30.477 m, resulting in an average radius of 25.397 m. The porosities of both the meter-scale and the kilometer-scale particle models are 0.17, while the density and friction coefficient of the particle material are 2500 kg/m^3 and 0.5 respectively. Because the geometrical similarity ratio (i.e. 1/1000) of the two particle models is equal to the corresponding particle diameter similarity ratio, the first similarity criterion of the upscale theory is satisfied between these two similar particle models. Since the same value of the contact stiffness is used in both the meter-scale

8.4 Test and Application Examples of the Particle Simulation Method

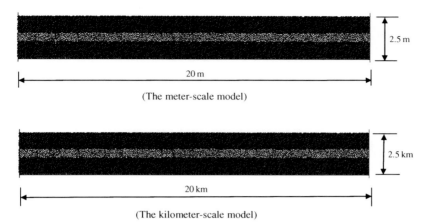

Fig. 8.16 Geometries of three-layer models of different length-scales: the middle layer thickness is 0.5 m for the meter-scale model, while it is 500 m for the kilometer-scale model

and kilometer-scale models, the second similarity criterion is obviously satisfied between these two models. The macroscopic tensile strength of the particle material is 80 MPa, while the macroscopic shear strength of the particle material is 800 MPa for both the meter-scale and the kilometer-scale models. As mentioned previously, if $\alpha = 1$, the values of the unit normal and tangential contact bond strengths are equal to those of the macroscopic tensile and shear strengths of the particle material (Zhao et al. 2007b). Because the normal and tangential contact bond strengths are directly proportional to the particle diameter, the third similarity criterion is also satisfied between these two models. To satisfy the fourth similarity criterion of the proposed upscale theory, the gravity acceleration of the kilometer-scale model is equal to 9.81 m/s^2, while the gravity acceleration of the meter-scale model is equal to 9810 m/s^2, implying that a gravity-acceleration similarity ratio of the meter-scale model to the kilometer-scale one is equal to 1000, which is the inverse of the geometrical similarity of these two similar particle models. Thus, the fourth similarity criterion as indicated by Eq. (8.49) is satisfied between the meter-scale and the kilometer-scale particle models.

Figure 8.17 shows the evolution of the folding process of the kilometer-scale model, in which the stiffer middle layer is embedded between softer upper and lower layers, while Fig. 8.18 shows a similar evolution of the folding process of the meter-scale model for several different deformation stages of the model. Note that brown segments are used to show crack patterns in these two figures. Since crack initiation and generation can be simulated in the particle simulation of both the kilometer-scale and meter-scale models, the corresponding crack patterns are also shown by brown segments in these two figures. It is clear that because the middle layer is 10 times stiffer than both the upper and the lower layers, the generated cracks are almost entirely located within this stiffer middle layer. In the case of horizontal shortening deformation equal to 10.8%, the first major crack occurs at the

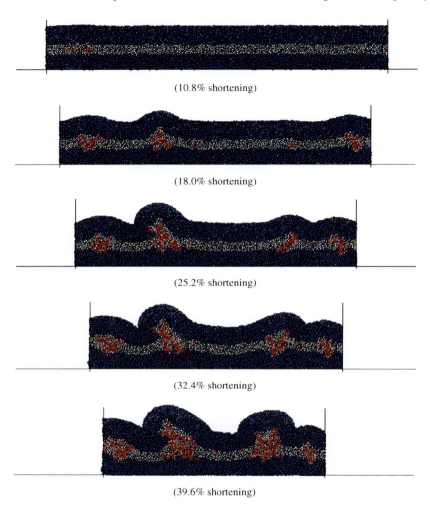

(10.8% shortening)

(18.0% shortening)

(25.2% shortening)

(32.4% shortening)

(39.6% shortening)

Fig. 8.17 Evolution of the folding process of a stiffer middle layer embedded by softer upper and lower layers (The kilometer-scale model)

left-hand side of both the kilometer-scale and the meter-scale models, indicating that the mechanical responses of both the models are identical within the elastic range of the particle material. Although there are some discrepancies between the simulation results of the two models at post-failure stages, the overall deformation patterns are very similar between the two similar models of different length-scales, indicating that the proposed upscale theory is correct and useful for revealing the intrinsic relationship between the two similar particle models of different length-scales, even though the two similar particle models are comprised of heterogeneous material

8.4 Test and Application Examples of the Particle Simulation Method 215

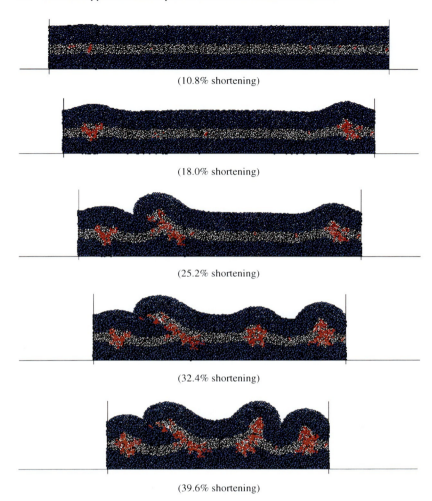

(10.8% shortening)

(18.0% shortening)

(25.2% shortening)

(32.4% shortening)

(39.6% shortening)

Fig. 8.18 Evolution of the folding process of a stiffer middle layer embedded by softer upper and lower layers (The meter-scale model)

regions. This implies that if the four similar criteria of the proposed upscale theory are satisfied, it is possible to use the mechanical response of a small length-scale model such as a laboratory length-scale model to investigate the potential mechanical response of a large length-scale model such as a geological length-scale model. From this point of view, the proposed upscale theory of the particle simulation provides a useful bridge between the simulation results obtained from any two similar particle models of different length-scales.

8.4.4 Particle Simulation of the Faulting Process Using the Proposed Particle Method

The problem associated with the crustal fault-propagation folding above rigid basement blocks is considered to illustrate the particle simulation of spontaneous crack generation problems in large-scale quasi-static geological systems. Figure 8.19 shows the geometry of the computational model, in which the length and height are 10 km and 2.5 km respectively. The dip angle of an underlying fault in the rigid basement is 60 degrees (i.e. $\theta = 60°$). The model is simulated by 4000 particles. The maximum and minimum radii of particles are approximately 30.48 m and 20.32 m, resulting in an average radius of 25.4 m. Although the model is mechanically comprised of one homogeneous layer, we use 10 approximately flat-lying and constant-thickness marker beds to monitor the deformation patterns. The macroscopic elastic modulus of the particle material used in the model is 5 GPa, resulting in a contact stiffness (in both the normal and the tangential directions) of 10 GN/m for the computational model. The normal/tangential contact bond strength of a particle is considered in the exactly same manner as in the previous section. This means that the value of the unit normal/tangential contact bond strength is equal to that of the macroscopic tensile/shear strength of the particle material, while the value of the normal/tangential contact bond strength of a particle is equal to the product of the unit normal/tangential contact bond strength and the diameter value of the particle. In this way, the variation of the normal/tangential contact bond strength of a particle with its diameter value is considered in the particle simulation. The macroscopic tensile strength of the particle material is 20 MPa, while the macroscopic shear strength of the particle material is 200 MPa for the computational model. The friction coefficient of the particle material is 0.5 and the density of the particle material is assumed to be 2500 kg/m^3 in the particle simulation. The loading period is 10 time-steps, while the frozen period is 1000 time-steps in the numerical simulation. Particles in contact with both the lateral boundaries are not allowed to move in the horizontal direction but are allowed to move in the vertical direction. Since the top of the computational model is a free surface, a stress-free boundary condition is applied to this boundary. The model is run to reach an initial equilibrium state due

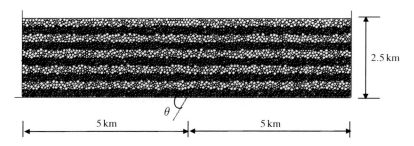

Fig. 8.19 Geometry of the computational model

8.4 Test and Application Examples of the Particle Simulation Method   217

to gravity. In order to simulate the slip of the underlying fault, the right half of the bottom is fixed, while the left half of the bottom is allowed to move in the direction that is parallel to the underlying fault plane in the rigid basement.

As we mentioned in the previous section, the second numerical simulation issue associated with the distinct element method is an inherent issue, which is caused by using the explicit dynamic relaxation method to solve a quasi-static problem. Although the problem related to this issue cannot be completely solved at this stage, an expedient measure is strongly recommended to carry out a particle-size sensitivity analysis of at least two different models, which have the same initial geometry but different total numbers of particles, to confirm the particle simulation result of a large-scale quasi-static system. For this purpose, the same problem as considered above is simulated using 8000 particles, so that the total number of particles used in this simulation is twice that used in the previous simulation. For ease of discussion, the previous model of 4000 particles is called the 4000-particle model, while the model of 8000 particles is defined as the 8000-particle model. For the 8000-particle model, the maximum and minimum radii of particles are approximately 21.55 m and 14.37 m, resulting in an average radius of 17.96 m. Note that the average radius of particles used in the 4000-particle model is 25.4 m.

It needs to be pointed out that in theory, the smallest particle size of a particle model is related directly to the material fracture toughness (Potyondy and Cundall 2004), especially under mixed compressive-extensile conditions. In the case of modeling damage processes for which macroscopic cracks form, the smallest particle size and model properties should be chosen to match the material fracture toughness as well as the unconfined compressive strength. However, it was also found that the formation of a failure plane and secondary macro-cracks may be independent of particle size under mixed compressive-shear conditions (Potyondy and Cundall 2004), which are those that we consider in this study. Nevertheless, in order to test whether or not the formation of macroscopic cracks is dependent on the smallest particle size, it is recommended that a particle-size sensitivity analysis of at least two different models, which have the same geometry but different smallest particle sizes, be carried out to confirm the particle simulation result of a large-scale quasi-static system.

Figure 8.20 shows a comparison of crack patterns within the 4000-particle and 8000-particle models respectively. Note that brown segments are used to show crack patterns in this figure and the forthcoming figure (i.e. Fig. 8.21). It is observed that in terms of the two major macroscopic cracks, both the 4000-particle model and the 8000-particle model produce the identical results, although the simulation result of the 8000-particle model is of higher resolution. This confirms that the particle simulation results obtained from the 4000-particle model is appropriate for showing the major macroscopic cracks in the computational model. It is also noted that the deformation pattern displayed in Fig. 8.20 is very similar to that reported in a previous publication (Finch et al. 2003). This demonstrates that in addition to the conceptual soundness, the proposed loading procedure is correct and useful for dealing with the numerical simulation of the brittle behavior of crustal rocks. For the above-mentioned reasons, the 8000-particle model is used hereafter to investigate

Fig. 8.20 Effect of the total number of particles on crack patterns in two computational models of the same initial geometry

(4000-particle model with showing particles)

(4000-particle model without showing particles)

(8000-particle model with showing particles)

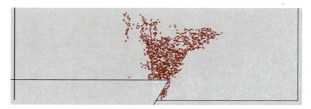

(8000-particle model without showing particles)

the effects of the dip angle of the underlying fault on the spontaneous crack generation patterns due to the crustal fault-propagation folding above rigid basement blocks.

Figure 8.20 also shows that spontaneous cracks are only generated in the central part of the computational model. Since most parts of the computational model are still in the elastic response state, they can be simulated more efficiently using continuum-mechanics-based numerical methods such as the finite element and finite difference methods (Zhao et al. 1999f). In this regard, it is not the most efficient process to use particles to simulate the whole computational model. To overcome this disadvantage, the combined use of both the continuum-mechanics-based

8.4 Test and Application Examples of the Particle Simulation Method

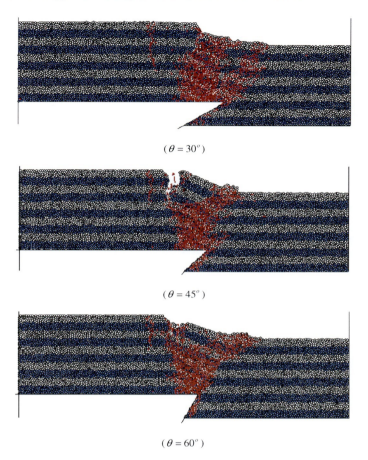

($\theta = 30^\circ$)

($\theta = 45^\circ$)

($\theta = 60^\circ$)

Fig. 8.21 Effect of the dip angle of the underlying fault on crack generation within the computational model (Crack pattern with particles)

numerical method and the particle simulation method has been proposed in recent years (Potyondy and Cundall 2004, Suiker and Fleck 2004, Fleck and Willis 2004). Since the continuum-mechanics-based numerical method and particle simulation method are used to simulate the elastic region and cracking region of a computational model respectively, the efficiency of the numerical simulation, as a whole, can be greatly improved. Since the main purpose of this investigation is to extend the application range of the particle simulation method from a laboratory scale into a geological scale, the particular issue of combining the particle simulation method with the continuum-mechanics-based numerical method is not discussed in detail here, for the sake of saving space.

Keeping the initial geometry and material properties of the computational model unchanged, three different dip angles, namely $\theta = 30°$, $\theta = 45°$ and $\theta = 60°$, of the underlying fault in rigid basement blocks are considered to investigate the effect of the dip angle of the underlying fault on spontaneous crack generation patterns in crustal brittle rocks. Figure 8.21 shows the related numerical simulation results of crack generation and deformation patterns in the computational model due to three different dip angles of the underlying fault. The results shown in this figure are obtained when the vertical displacement of the left-hand-side basement is about 30% of the whole thickness of the computational model. It is observed from these simulation results that there are some remarkable differences in the crack generation and deformation patterns due to these three different dip angles, implying that the dip angle of the underlying fault has a significant influence on spontaneous crack generation patterns in crustal brittle rocks. The major reason for causing such remarkable differences in the simulation results is that for the same amount of vertical displacement of the left-hand-side basement, the corresponding horizontal displacement of the left-hand-side basement is significant different for the three different dip angles of the underlying fault. For example, when the vertical displacement of the left-hand-side basement is about 30% of the whole thickness of the computational model, the corresponding horizontal displacement of the left-hand-side basement is about 51.97, 30 and 17.32% of the whole thickness of the computational model in the case of the dip angle being $30°$, $45°$ and $60°$ respectively. Due to the significant difference in the horizontal displacement of the left-hand-side basement, the crack pattern in the computational model of $\theta = 30°$ is more diffuse than that in the computational model of $\theta = 60°$. In the case of $\theta = 60°$, a macroscopic crack (i.e. a new fault) is formed in the slip direction of the underlying fault and also passes the tip point of the left-hand-side basement. However, in the cases of $\theta = 30°$ and $\theta = 45°$, the dip angle of the resulting new fault in the computational model is steeper than that of the underlying fault in the rigid basement blocks. In addition, it is noted that in the case of $\theta = 45°$, there is a large crack within the hanging-wall of the computational model. This crack penetrates into about 40% of the whole thickness of the computational model. This phenomenon further demonstrates that the dip angle of the underlying fault has a significant effect on spontaneous crack generation patterns in crustal brittle rocks.

Summary Statements

As an amalgamation of traditional geoscience and contemporary computational science, computational geoscience has become an emerging discipline in the past decade. On the one hand, computational geoscience provides geoscientists with useful modern scientific tools for revealing dynamic processes and mechanisms behind the field observations of complicated and complex geoscience phenomena, but on the other hand, it provides computational scientists with challenging problems of large scales and multiple processes. In addition to the well-known experimental and theoretical analysis methods, computational simulation has become a third dominating and indispensable scientific method for solving a broad range of scientific and engineering problems. Without computational simulation, it is impossible to solve geoscience problems at a high lever of understanding the intrinsic dynamic processes and mechanisms that lead to the oberserved geoscience phenomena in nature.

Owing to the broad and diversity characteristics of geoscience problems, a typical kind of geoscience problem, known as ore body formation and mineralization in hydrothermal systems within the upper crust of the Earth, has been considered to deal with the computational aspects of recent developments of computational geoscience. Toward this end, advanced numerical methods, procedures and algorithms have been systematically presented in this monograph. Through applying these numerical methods, procedures and algorithms to some specific aspects of ore body formation and mineralization, the following conclusions have been drawn from the research work reported in the monograph.

(1) The newly-developed computational geoscience discipline is obviously of multi-disciplinary nature crossing many fields of science such as mathematics, physics, chemistry, computational science, geoscience and so forth. The ultimate aim of computational geoscience is to deal with the origin, history and behaviours of the Earth system in a scientific and predictive manner. The research methodology of computational geoscience is a comprehensive research methodology, which is formed by combining field observation, theoretical analysis, numerical simulation and field validation. The primary aim of using this research methodology is to investigate the dynamic processes and mechanisms

of observed geoscience phenomena, rather than to describe the observed phenomena themselves.

(2) Convective pore-fluid flow, known as the steady-state Horton-Rogers-Lapwood problem, is an important mechanism to control ore body formation and mineralization in hydrothermal systems within the upper crust of the Earth. This kind of problem belongs to a kind of bifurcation problem, from a nonlinear mathematics point of view. A progressive asymptotic approach procedure has been presented for solving the steady-state Horton-Rogers-Lapwood problem in a fluid-saturated porous medium. This problem possesses a bifurcation and therefore makes the direct use of conventional finite element methods difficult. Even if the Rayleigh number is high enough to drive the occurrence of natural convection in a fluid-saturated porous medium, the conventional methods often produce a trivial non-convective solution. This difficulty can be overcome using the progressive asymptotic approach procedure associated with the finite element method. The method considers a series of modified Horton-Rogers-Lapwood problems in which gravity is assumed to tilt a small angle away from vertical. The main idea behind the progressive asymptotic approach procedure is that through solving a sequence of such modified problems with decreasing tilt, an accurate non-zero velocity solution to the Horton-Rogers-Lapwood problem can be obtained. This solution provides a very good initial prediction for the solution to the original Horton-Rogers-Lapwood problem so that the non-zero velocity solution can be successfully obtained when the tilted angle is set to zero. Comparison of numerical solutions with analytical ones to a benchmark problem of any rectangular geometry has demonstrated the usefulness of the proposed progressive asymptotic approach procedure for dealing with convective pore-fluid flow problems within the upper crust of the Earth.

(3) To deal with coupled problem between material deformation, pore-fluid flow, heat transfer, mass transport and chemical reactions in hydrothermal systems within the upper crust of the Earth, the combined use of two or more commercially available computer codes is a favorable choice. A consistent point-searching algorithm for solution interpolation in unstructured meshes consisting of 4-node bilinear quadrilateral elements has been presented to translate and transfer solution data between two totally different meshes that are used in two different computer codes, both commercially available. The proposed algorithm has the following significant advantages: first, the use of a point-searching strategy allows a point in one mesh to be accurately related to an element (containing this point) in another mesh. Thus, to translate/transfer the solution of any particular point from the mesh used in one computer code to that in another computer code, only one element needs to be inversely mapped. This certainly minimizes the number of elements, to which the inverse mapping is applied, so that the present algorithm is very effective and efficient. Second, analytical solutions to the local coordinates of any point in a four-node quadrilateral element, which are derived in a rigorous mathematical manner, make it possible to carry out an inverse mapping process very effectively and

efficiently. Third, the use of consistent interpolation enables the interpolated solution to be compatible with an original solution and therefore guarantees the interpolated solution of extremely high accuracy. The related results from the test problem have demonstrated the generality, accuracy, effectiveness, efficiency and robustness of the proposed consistent point-searching integration algorithm.

(4) To effectively and efficiently use the finite element method for solving fluid-rock interaction problems of subcritical Zhao numbers in pore-fluid saturated hydrothermal/sedimentary basins, A term splitting algorithm on the basis of a new concept of the generalized concentration of a solid mineral has been presented to deal with the following three fundamental issues associated with the fluid-rock interaction problems. First, since the fluid-rock interaction problem involves heterogeneous chemical reactions between reactive aqueous chemical species in the pore-fluid and solid minerals in the rock masses, it is necessary to develop a new concept of the generalized concentration of a solid mineral, so that two types of reactive mass transport equations, namely the conventional mass transport equation for the aqueous chemical species in the pore-fluid and the degenerated mass transport equation for the solid minerals in the rock mass, can be solved simultaneously in computation. Second, because the reaction area between the pore-fluid and mineral surfaces is basically a function of the generalized concentration of the solid mineral, there is a definite need to appropriately consider the dependence of the dissolution rate of a dissolving mineral on its generalized concentration in the numerical analysis. Third, to consider porosity evolution with time in the transient analysis of fluid-rock interaction problems, the concept of the equivalent source/sink terms in mass transport equations needs to be developed to convert the problem of variable mesh Peclet number and Courant number into the problem of constant mesh Peclet and Courant numbers. The related numerical results have demonstrated the usefulness and robustness of the proposed term splitting algorithm for solving fluid-rock interaction problems of subcritical Zhao numbers in pore-fluid saturated hydrothermal and sedimentary basins.

(5) The chemical-dissolution-front propagation problem exists ubiquitously not only in ore forming systems within the upper crust of the Earth, but also in many other scientific and engineering fields. To solve this problem, it is necessary to deal with a coupled system between porosity, pore-fluid pressure and reactive chemical-species transport in fluid-saturated porous media. Due to the morphological instability of a chemical dissolution front, this problem needs to be solved numerically. A segregated algorithm on the basis of a combination of the finite element and finite difference methods has been proposed for simulating the morphological evolution of chemical dissolution fronts in reactive transport systems of critical and supercritical Zhao numbers. A set of analytical solutions have been derived for a benchmark problem to verify the proposed numerical procedure. Not only can the derived analytical solutions be used to verify any numerical method before it is used to solve this kind of chemical-dissolution-front propagation problem, but also they can be used

to understand the fundamental mechanisms behind the morphological instability of a chemical dissolution front during its propagation within fluid-saturated porous media of critical and supercritical Zhao numbers. The related numerical results have demonstrated that the proposed segregated algorithm and the related numerical procedure are useful for and capable of simulating the morphological instability of a chemical dissolution front within the fluid-saturated porous medium.

(6) Non-equilibrium redox chemical reactions of high orders are ubiquitous in fluid-saturated porous rocks within the crust of the Earth. The numerical modelling of such high-order chemical reactions becomes a challenging problem because these chemical reactions are not only produced strong nonlinear source/sink terms for reactive transport equations, but also often coupled with the fluids mixing, heat transfer and reactive mass transport processes. In order to solve this problem effectively and efficiently, it is desirable to reduce the total number of reactive transport equations with strong nonlinear source/sink terms to a minimum in a computational model. For this purpose, a decoupling procedure on the basis of the concept of the chemical reaction rate invariant has been developed for dealing with fluids mixing, heat transfer and non-equilibrium redox chemical reactions in fluid-saturated porous rocks. Using the proposed decoupling procedure, only one reactive transport equation, which is used to describe the distribution of the chemical product and has a strong nonlinear source/sink term, needs to be solved for each of the non-equilibrium redox chemical reactions. The original reactive transport equations of the chemical reactants with strong nonlinear source/sink terms are turned into the conventional mass transport equations of the chemical reaction rate invariants without any nonlinear source/sink terms. A testing example, for some aspects of which the analytical solutions are available, is used to verify the proposed numerical procedure. The proposed decoupling procedure associated with the finite element method has been used to investigate mineral precipitation patterns due to two reactive fluids focusing and mixing within permeable faults within the upper crust of the Earth. The related numerical solutions have demonstrated that the proposed numerical procedure is useful and applicable for dealing with the coupled problem between fluids mixing, heat transfer and non-equilibrium redox chemical reactions of high orders in fluid-saturated porous rocks.

(7) The solidification of intruded magma in porous rocks can result in the following two consequences: (1) heat release due to the solidification of the interface between the rock and intruded magma and (2) mass release of the volatile fluids in the region where the intruded magma is solidified into the rock. Traditionally, the intruded magma solidification problem is treated as a moving interface (i.e., the solidification interface between the rock and intruded magma) problem to consider these consequences in conventional numerical methods. An equivalent source algorithm has been presented to simulate thermal and chemical consequences/effects of magma intrusion in geological systems, which are composed of porous rocks. Using the proposed equivalent source algorithm, an original

magma solidification problem with a moving boundary between the rock and intruded magma has been transformed into a new problem without the moving boundary but with the proposed mass sources and physically equivalent heat sources. The major advantage in using the proposed equivalent source algorithm is that a fixed mesh of finite elements with a variable integration timestep can be employed to simulate the consequences and effects of the intruded magma solidification using the conventional finite element method. The related results from a benchmark magma solidification problem have demonstrated the correctness and usefulness of the proposed equivalent source algorithm.

(8) To extend the application range of the particle simulation method from a laboratory scale to a large scale such as a geological scale, we need to deal with an upscale issue associated with simulating spontaneous crack generation problems in large-scale quasi-static systems. Toward this direction, three important simulation issues, which may affect the quality of the particle simulation results of a quasi-static system, have been addressed. The first simulation issue is how to determine the particle-scale mechanical properties of a particle from the measured macroscopic mechanical properties of rocks. The second simulation issue is that fictitious time, rather than physical time, is used in the particle simulation of a quasi-static problem. The third simulation issue is that the conventional loading procedure used in the particle simulation method is conceptually inaccurate, at least from the force propagation point of view. A new loading procedure and an upscale theory have been presented to solve the conceptual problems arising from the first and third simulation issues. The proposed loading procedure is comprised of two main types of periods, a loading period and a frozen period. Using the proposed loading procedure and upscale theory, the parameter selection problem stemming from the first issue can be solved. Since the second issue is an inherent one, it is strongly recommended that a particle-size sensitivity analysis of at least two different models, which have the same geometry but different smallest particle sizes, be carried out to confirm the particle simulation result of a large-scale quasi-static system. The related simulation results have demonstrated the usefulness and correctness of the proposed loading procedure and upscale theory for dealing with spontaneous crack generation problems in large-scale quasi-static geological systems.

(9) This is the end of this monograph, but just the beginning of the computational geoscience world in the sense that more and more complicated and complex geoscience problems need to be solved from now on. During writing this monograph, there was the most disastrous earthquake occurring in Wenchuan, China. Unfortunately, both the time and location for the occurrence of this earthquake cannot be predicted by the present day's knowledge of geoscientists. While we express our deep condolence to those who lost their precious lives during this earthquake, we really hope that with the further development of computational geoscience, we can understand all the dynamic processes and mechanisms that control the occurrence of earthquakes better, so that we could accurately predict it in such a way as predicting weather today.

References

Alavyoon, F., 1993. On natural convection in vertical porous enclosures due to prescribed fluxes of heat and mass at the vertical boundaries, *International Journal of Heat and Mass Transfer*, **36**, 2479–2498.

Alexiades, V. and Solomon, A. D., 1993. *Mathematical Modelling of Melting and Freezing Processes*, Hemisphere Publishing Co., Washington, D. C.

Antonellini, M. A. and Pollard, D. D., 1995. Distinct element modeling of deformation bands in sandstone, *Journal of Structural Goelogy*, **17**, 1165–1182.

Appold, M. S. and Garven, G., 2000. Reactive flow models of ore formation in the Southeast Missouri district, *Economic Geology and the Bulletin of the Society of Economic Geologists*, **95**, 1605–1626.

Bardet, J. P. and Proubet, J., 1992. The structure of shear bands in idealized granular materials, *Applied Mechanics Review*, **45**, 118–122.

Barns, H. L., 1997., *Geochemistry of Hydrothermal Ore Deposits*, John Wiley and Sons Inc., New York.

Barsoum, R. S., 1976. On the use of isoparametric finite elements in linear fracture mechanics, *International Journal for Numerical Methods in Engineering*, **10**, 25–37.

Barsoum, R. S., 1977. Triangular quarter point elements elastic and perfectly plastic crack tip elements, *International Journal for Numerical Methods in Engineering*, **11**, 85–98.

Bear, J., 1972. *Dynamics of Fluids in Porous Media*, American Elsevier Publishing Company, New York.

Bear, J. and Bachmat, Y., 1990. *Introduction to Modelling of Transport Phenomena in Porous Fractured Media*. Kluwer Academic Press.

Beyer, R. P. and LeVeque, R. J., 1992. Analysis of a one-dimensional model for the immersed boundary method, *SIAM Journal on Numerical Analysis*, **29**, 332–364.

Bons, P. D., Dougherty-Page, J. and Elburg, M. A., 2001. Stepwise accumulation and ascent of magmas, *Journal of Metamorphic Geology*, **19**, 627–633.

Bouchard, P. O., Bay, Y. and Chastel Y., 2003. Numerical modelling of crack propagation: automatic remeshing and comparison of different criteria, *Computer Methods in Applied Mechanics and Engineering*, **192**, 3887–3908.

Braun, J., Munroe, S. M. and Cox, S. F., 2003. Transient fluid flow in and around a fault, *Geofluids*, **3**, 81–87.

Buck, W. R. and Parmentier, E. M., 1986. Convection between young oceanic lithosphere: Implication for thermal structure and gravity, *Journal of Geophysical Research*, **91**, 1961–1974.

Burbidge, D. R. and Braun, J., 2002. Numerical models of the evolution of accretionary wedges and fold-and-thrust belts using the distinct-element method, *Geophysical Journal International*, **148**, 542–561.

Buretta, R. J. and Berman, A. S., 1976. Convective heat transfer in a liquid saturated porous layer, *ASME Journal of Applied Mechanics*, **43**, 249–253.

Burnham, C. W., 1979. The importance of volatile constituents, in *The Evolution of the Igneous Rocks: Fiftieth Anniversary Perspectives*, Princeton University Press, New York.

Caltagirone, J. P., 1975. Thermoconvective instabilities in a horizontal layer, *Journal of Fluid Mechanics*, **72**, 269–287.

Caltagirone, J. P., 1976. Thermoconvective instabilities in a porous medium bounded by two concentric horizontal cylinders, *Journal of Fluid Mechanics*, **76**, 337–362.

Camborde, F., Mariotti, C. and Donze, F. V., 2000. Numerical study of rock and concrete behavior by distinct element modeling, *Computers and Geotechnics*, **27**, 225–247.

Carslaw, H. and Jaeger, J., 1959. *Conduction of Heat in Solids*, Clarendon Press, Oxford.

Chadam, J., Hoff, D., Merino, E., Ortoleva, P. and Sen, A., 1986. Reactive infiltration instabilities, *IMA Journal of Applied Mathematics*, **36**, 207–221.

Chadam, J., Ortoleva, P. and Sen, A., 1988. A weekly nonlinear stability analysis of the reactive infiltration interface, *IMA Journal of Applied Mathematics*, **48**, 1362–1378.

Chen, J. S. and Liu, C. W., 2002. Numerical simulation of the evolution of aquifer porosity and species concentrations during reactive transport, *Computers and Geosciences*, **28**, 485–499.

Chevalier, S., Bernard, D. and Joly, N., 1999. Natural convection in a porous layer bounded by impervious domains: from numerical approaches to experimental realization, *International Journal of Heat and Mass Transfer*, **42**, 581–597.

Cline, J. S. and Hofstra, A. A., 2000. Ore-fluid evolution at the Getchell Carline-type gold deposit, Nevada, USA, *European Journal of Mineralogy*, **12**, 195–212.

Combarnous, M. A. and Bories, S. A., 1975. Hydrothermal convection in saturated porous media, *Adv. Hydroscience*, **10**, 231–307.

Connolly, J. A. D., 1997. Mid-crustal focused fluid movements: Thermal consequences and silica transport. In: Jamtveit, B. and Yardley, B. W. D. (Eds.), *Fluid Flow and Transport in Rocks: Mechanics and Effects*, Chapman and Hall, London, pp. 235–250.

Cook, R. D., Malkus, D. S. and Plesha, M. E., 1989. *Concepts and Applications of Finite Element Analysis*, Wiley, New York.

Cox, S. F., Sun, S. S., Etheridge, M. A., Wall, V. J. and Potter, T. F., 1995. Structural and geochemical controls on the development of turbidite-hosted gold quartz vein deposits, Wattle Gully mine, central Victoria, Australia, *Economic Geology and the Bulletin of the Society of Economic Geologists*, **90**, 1722–1746.

Crank, J., 1984. *Free and Moving Boundary Problems*, Oxford University Press, London.

Cundall, P. A. and Strack, O. D. L., 1979. A discrete numerical model for granular assemblies, *Geotechnique*, **29**, 47–65.

Cundall, P. A., 2001. A discontinuous future for numerical modelling in geomechanics?, *Proceedings of the Institution of Civil Engineers: Geotechnical Engineering*, **149**, 41–47.

Detournay, E. and Cheng, A. H. D., 1993. Fundamentals of poroelasticity. In: J. A. Hudson and C. Fairhurst (Eds.), *Comprehensive Rock Engineering*, **Vol.2**: Analysis and Design Methods, Pergamon Press, New York.

Doin, M. P., Fleitout, L. and Christensen, U., 1997. Mantle convection and stability of depleted and undepleted continental lithosphere, *Journal of Geophysical Research*, **102**, 2771–2787.

Donze, F., Mora, P. and Magnier, S. A., 1994. Numerical simulation of faults and shear zones, *Geophysical Journal International*, **116**, 46–52.

Donze, F., Magnier, S. A. and Bouchez, J., 1996. Numerical modeling of a highly explosive source in an elastic-brittle rock mass, *Journal of Geophysical Research*, **101**, 3103–3112.

Erdogan, F. and Sih, G. C., 1963. On the crack extension in plane loading and transverse shear, *Journal Basic Engineering*, **85**, 519–527.

Everett, C. E., Wilkinson, J. J. and Rye, D. M., 1999. Fracture-controlled fluid flow in the Lower Palaeozoic basement rocks of Ireland: implications for the genesis of Irish-type Zn-Pb deposits. In: McCaffrey, K. J. W., Lonergan. L. and Wilkinson, J. J. (Eds), *Fractures, Fluid Flow and Mineralization*, Geological Society, London, Special Publications **155**, 247–276.

References

Finch, E., Hardy, S. and Gawthorpe, R., 2003. Discrete element modeling of contractional fault-propagation folding above rigid basement fault blocks, *Journal of Strctural Geology*, **25**, 515–528.

Finch, E., Hardy, S. and Gawthorpe, R., 2004. Discrete element modeling of extensional fault-propagation folding above rigid basement fault blocks, *Basin Research*, **16**, 489–506.

Fleck, N. A. and Willis, J. R., 2004. Bounds and estimates for the effects of strain gradients upon the effective plastic properties of an isotropic two-phase composite, *Journal of the Mechanics and Physics of Solids*, **52**, 1855–1888.

Fluid Dynamics International, 1997. *Fluid Dynamics Analysis Package: FIDAP*, Fluid Dynamics International Inc., Illinois.

G+D Computing, 1991. *Strand6 Reference Manual and User Guide*, G+D Computing, Sydney.

Garven, G. and Freeze, R. A., 1984. Theoretical analysis of the role of groundwater flow in the genesis of stratabound ore deposits: Mathematical and numerical model, *American Journal of Science*, **284**, 1085–1124.

Garven, G., Appold, M. S., Topygina, V. I. and Hazlett, T. J., 1999. Hydrologic modeling of the genesis of carbonate-hosted lead-zinc ores. *Hydrogeology Journal*, **7**, 108–126.

Gawin, D., Pesavento, F. and Schrefler, B. A., 2003. Modelling of hygro-thermal behaviour of concrete at high temperature with thermal-chemical and mechanical material degradation, *Computer Methods in Applied Mechanics and Engineering*, **192**, 1731–1771.

Gobin, D. and Bennacer, R., 1994. Double diffusion convection in a vertical fluid layer: onset of the convection regime, *Physical Fluids*, **6**, 59–67.

Gow, P., Upton, P., Zhao, C. and Hill, K., 2002. Copper-gold mineralization in the New Guinea: Numerical modeling of collision, fluid flow and intrusion-related hydrothermal systems, *Australian Journal of Earth Sciences*, **49**, 753–771.

Goyeau, B., Songbe, J. P. and Gobin, D., 1996. Numerical study of double-diffusive convection in a porous cavity using Darcy-Brinkman formulation, *International Journal of Heat and Mass Transfer*, **39**, 1363–1378.

Haar L., Gallagher, J. S. and Kell, G. S., 1984. NBS/NRC Steam Tables: Thermodynamic and Transport Properties and Computer Programs for Vapor and Liquid States of Water in SI Units, Taylor & Francis, New York.

Hellen, T. K., 1975. On the method of virtural crack extensions, *International Journal for Numerical Methods in Engineering*, **9**, 187–207.

Hitzmann, M. W., 1995. Mineralisation in the Irish Zn-Pb-(Ba-Ag) Orefield. In: Anderson, K., Ashton, J., Earls, G., Hitzman, M. and Tear, S. (Eds.), *Irish Carbonate-Hosted Deposits*, Society of Economic Geologists, Guidebook Series, **21**, 25–61.

Hobbs, B. E., Zhang, Y., Ord, A. and Zhao, C., 2000. Application of coupled deformation, fluid flow, thermal and chemical modelling to predictive mineral exploration, *Journal of Geochemical Exploration*, **69**, 505–509.

Horton, C. W. and Rogers, F. T., 1945. Convection currents in a porous medium, *Journal of Applied Physics*, **16**, 367–370.

Imber, J., Tuckwell, G. W., Childs, C., Walsh, J. J., Manzocchi, T., Heath, A. E., Bonson, C. G. and Strand, J., 2004. Three-dimensional distinct element modeling of relay growth and breaching along normal faults, *Journal of Structural Geology*, **26**, 1897–1911.

Ingraffea, A. R. and Manu, C., 1980. Stress-intensity factor computation in three dimensions with quarter-point elements, *International Journal for Numerical Methods in Engineering*, **15**, 1427–1445.

Islam, M. R. and Nandakumar, K., 1990. Transient convection in saturated porous layers, *International Journal of Heat Mass Transfer*, **33**, 151–161.

Itasca Consulting Group, 1995. *Fast Lagrangian Analysis of Continua (FLAC)*, Itasca Consulting Group, Inc., Minnesota.

Itasca Consulting Group, Inc., 1999. *Particle Flow Code in Two Dimensions (PFC2D)*, Minneapolis, Minnesota, USA.

Iwashita, K. and Oda, M., 2000. Micro-deformation mechanism of shear banding process based on modified distinct element method, *Powder Technology*, **109**, 192–205.

Jaeger, J. C. and Cook, N. G. W., 1976. *Fundamentals of rock mechanics*, Chapman & Hall, London.

Jiang, Z., Oliver, N. H. S., Barr, T. D., Power, W. L. and Ord, A., 1997. Numerical modelling of fault-controlled fluid flow in the genesis of tin deposits of the Malage ore field, Gejiu mining district, China, *Economic Geology*, **92**, 228–247.

Johnson, A. M. and Pollard, D. D., 1973. Mechanics of growth of some laccolithic intrusions in the Henry Mountains, Utah, I, *Tectonophysics*, **18**, 261–309.

Joly, N., Bernard, D. and Menegazzi, P., 1996. ST2D3D: A finite element program to compute stability criteria for natural convection in complex porous structures, *Numerical Heat Transfer, Part B*, **29**, 91–112.

Kaviany, M., 1984. Thermal convective instabilities in a porous medium, *ASME Journal of Heat Transfer*, **106**, 137–142.

Khoei, A. R. and Lewis, R. W., 1999. Adaptive finite element remeshing in a large deformation analysis of metal power forming, *International Journal for Numerical Methods in Engineering*, **45**, 801–820.

Klerck, P. A., Sellers, E. J. and Owen, D. R. J., 2004. Discrete fracture in quasi-brittle materials under compressive and tensile stress states, *Computer Methods in Applied Mechanics and Engineering*, **193**, 3035–3056.

Kwak, D. Y., Cheon, J. S. and Im, Y. T., 2002. Remeshing for metal forming simulations: part I: Two-dimensional quadrilateral remeshing, *International Journal for Numerical Methods in Engineering*, **53**, 2436–2500.

Lapwood, E. R., 1948. Convection of a fluid in a porous medium, *Proceedings of the Cambridge Philosophical Society*, **44**, 508–521.

Lasaga, A. C., 1984. Chemical kinetics of water-rock interactions, *Journal of Geophysical Research*, **89**, 4009–4025.

Lebon, G. and Cloot, A., 1986. A thermodynamical modeling of fluid flows through porous media: Application to natural convection, *International Journal of Heat Mass Transfer*, **29**, 381–390.

Lee, N. S. and Bathe, K. J., 1994. Error indicators and adaptive remeshing in large deformation finite element analysis, *Finite Elements in Analysis and Design*, **16**, 99–139.

Lewis, R. W. and Schrefler, B. A., 1998. *The Finite Element Method in the Static and Dynamic Deformation and Consolidation of Porous Media*, John Wiley & Sons, New York.

Li, F. Z., Shih, C. F. and Needleman, A., 1985. A comparison of methods for calculating energy release rate, *Engineering Fracture Mechanics*, **21**, 405–421.

Lin, G., Zhao, C., Hobbs, B. E., Ord, A. and Muhlhaus, H. B., 2003. Theoretical and numerical analyses of convective instability in porous media with temperature–dependent viscosity, *Communications in Numerical Methods in Engineering*, **19**, 787–799.

Lister, J. R. and Kerr, R. C., 1991. Fluid-mechanics models of crack propagation and their application to magma transport in dykes, *Journal of Geophysical Research*, **96**, 10049–10077.

Liu, L., Yang, G., Peng, S. and Zhao, C., 2005. Numerical modelling of coupled geodynamical processes and its role in facilitating predictive ore discovery: An example from Tongling, China, *Resource Geology*, **55**, 21–31.

Lorenzi, H. G., 1985, Energy release rate calculations by the finite element method, *Engineering Fracture Mechanics*, **21**, 129–143.

Mamou, M., Vasseur, P. and Bilgen, E., 1998. Double-diffusive convection instability in a vertical porous enclosure, *Journal of Fluid Mechanics*, **368**, 263–289.

Marsh, B. D., 1982. On the mechanics of igneous diapirism, stoping, and zone melting, *American Journal of Science*, **282**, 808–855.

Matthai, S. K., Henley, R. W. and Heinrich, C. A., 1995. Gold precipitation by fluid mixing in bedding-parallel fractures near carbonaceous slates at the Cosmopolitan Howley gold deposit, northern Australia, *Economic Geology and the Bulletin of the Society of Economic Geologists*, **90**, 2123–2142.

Mazouchi, A. and Homsy, G. M., 2000. Thermocapillary migration of long bubbles in cylindrical capillary tubes, *Physics of Fluids*, **12**, 542–549.

References

McBride, A. Govender, I., Powell, M. and Cloete, T., 2004. Contributions to the experimental validation of the discrete element method applied to tumbling mills, *Engineering Computations*, **21**, 119–136.

McKibbin, R. and O'Sullivan, M. J., 1980. Onset of convection in a layered porous medium heated from below, *Journal of Fluid Mechanics*, **96**, 375–393.

McLellan, J, G., Oliver, N. H. S., Ord, A., Zhang, Y. and Schaubs, P. M., 2003. A numerical modelling approach to fluid flow in extensional environments: implications for genesis of large microplaty hermatite ores, *Journal of Geochemical Exploration*, **78–79**, 675–679.

Murphy, F. C., Ord, A., Hobbs, B. E., Willetts, G. and Barnicoat, A. C., 2008. Targeting stratiform Zn-Pb-Ag massive sulfide deposits in Ireland through numerical modeling of coupled deformation, thermal transport and fluid flow, *Economic Geology*, **103**, 1437–1458.

Nguyen, H. D., Pack, S. and Douglass, R. W., 1994. Study of double diffusive convection in a layered anisotropic porous medium, *Numerical Heat Transfer*, **26**, 489–505.

Nield, D. A., 1968. Onset of thermohaline convection in a porous medium, *Water Resources Research*, **11**, 553–560.

Nield, D. A. and Bejan, A., 1992. *Convection in Porous Media*, Springer-Verlag, New York.

Nithiarasu, P., Seetharamu, K. N. and Sundararajan, T., 1996. Double-diffusive natural convection in an enclosure filled with fluid-saturated porous medium: a generalized non-Darcy approach, *Numerical Heat Transfer*, **30**, 430–436.

Obdam, A. N. M. and Veling, E. J. M., 1987. Elliptical inhomogeneities in groundwater flow: An analytical description, *Journal of Hydrology*, **95**, 87–96.

Oliver, N. H. S., Pearson, P. J., Holcombe, R. J. and Ord, A., 1999. Mary Kathleen metamorphic-hydrothermal uranium-rare-earth element deposit: ore genesis and numerical model of coupled deformation and fluid flow, *Australian Journal of Earth Sciences*, **46**, 467–484.

Oliver N. H. S., 2001. Linking of regional and local hydrothermal systems in the mid-crust by shearing and faulting, *Tectonophysics*, **335**, 147–161.

Oliver, N. H. S., Ord, A., Valenta, R. and Upton, P., 2001. Deformation, fluid flow, and ore genesis in heterogeneous rocks, with examples and numerical models from the Mount Isa District, Australia, *Reviews in Economic Geology*, **14**, 51–74.

Ord, A. and Oliver, N. H. S., 1997. Mechanical controls on fluid flow during regional metamorphism: some numerical models, *Journal of Metamorphic Geology*, **15**, 345–359.

Ord, A., Hobbs, B. E., Zhang, Y., Broadbent, G. C., Brown, M., Willetts, G., Sorjonen-Ward, P., Walshe, J. and Zhao, C., 2002. Geodynamic modelling of the Century deposit, Mt Isa Province, Queensland, *Australian Journal of Earth Sciences*, **49**, 1011–1039.

Ord, A. and Sorjonen-Ward, P., 2003. Simulating the Outokumpu mineralising system, *The AusIMM Bulletin*, **5**, 46–47.

Ormond, A. and Ortoleva, P., 2000. Numerical modeling of reaction-induced cavities in a porous rock, *Journal of Geophysical Research*, **105**, 16737–16747.

Ortoleva, P., Chadam, J., Merino, E. and Sen, A., 1987. Geochemical self-organization II: The reactive-infiltration instability, *American Journal of Science*, **287**, 1008–1040.

Osher, S. and Sethian, J. A., 1998. Fronts propagating with curvature dependent speed: algorithms based on Hamilton-Jacobi formulations, *Journal of Computational Physics*, **79**, 12–49.

Owen, D. R. J., Feng, Y. T., de Souza Neto, E. A., Cottrell, M. G., Wang, F., Andrade Pires, F. M. and Yu, J., 2004. The modelling of multi-fracturing solids and particular media, *International Journal for Numerical Methods in Engineering*, **60**, 317–339.

Palm, E., Weber, J. E. and Kvernvold, O., 1972. On steady convection in a porous medium, *Journal of Fluid Mechanics*, **54**, 153–161.

Peacock, S. M., 1989. Numerical constraints on rates of metamorphism, fluid production and fluid flux during regional metamorphism, *Geological Society of America Bulletin*, **101**, 476–485.

Phillips, O. M., 1991. *Flow and Reactions in Permeable Rocks*, Cambridge University Press, Cambridge.

Pillatsis, G., Taslim, M. E. and Narusawa, U., 1987. Thermal instability of a fluid-saturated porous medium bounded by thin fluid layers, *ASME Journal of Heat Transfer*, **109**, 677–682.

Potyondy, D. O. and Cundall, P. A., 2004. A bonded-particle model for rock, *International Journal of Rock Mechanics and Mining Science*, **41**, 1329–1364.

Raffensperger, J. P. and Garven, G., 1995, The formation of unconformity-type uranium ore deposits: Coupled hydrochemical modelling, *American Journal of Science*, **295**, 639–696.

Rice, J. R., 1968. A path independent integral and the approximate analysis of strain concentrations by notches and cracks, *Journal of Applied Mechanics*, **35**, 379–386.

Riley, D. S. and Winters, K. H., 1989. Modal exchange mechanisms in Lapwood convection, *Journal of Fluid Mechanics*, **204**, 325–358.

Rubin, A. M., 1995. Propagation of magma-filled cracks, *Annual Reviews of Earth Planetary Science*, **23**, 287–336.

Saltzer, S. D. and Pollard, D. D., 1992. Distinct element modeling of structures formed in sedimentary overburden by extensional reactivation of basement normal faults, *Tectonics*, **11**, 165–174.

Salman, A. D. and Gorham, D. A., 2000. The fracture of glass spheres, *Powder Technology*, **107**, 179–185.

Schafer, D., Schafer, W. and Kinzelbach, W., 1998a. Simulation of reactive processes related to biodegradation in aquifers: 1. Structure of the three-dimensional reactive transport model, *Journal of Contaminant Hydrology*, **31**, 167–186.

Schafer, D., Schafer, W. and Kinzelbach, W., 1998b. Simulation of reactive processes related to biodegradation in aquifers: 2. Model application to a column study on organic carbon degradation, *Journal of Contaminant Hydrology*, **31**, 187–209.

Schaubs, P. and Zhao, C., 2002. Numerical modelling of gold-deposit formation in the Bendigo-Ballarat zone, Victoria, *Australian Journal of Earth Sciences*, **49**, 1077–1096.

Scheidegger, A. E., 1974. *The Physics of Flow through Porous Media*, University of Toronto Press, Toronto.

Schopfer, M. P. J., Childs, C. and Walsh, J. J., 2006. Localization of normal faults in multilayer sequences, *Journal of Structural Geology*, **28**, 816–833.

Schrefler, B. A., 2004. Multiphase flow in deforming porous material, *International Journal for Numerical Methods in Engineering*, **60**, 27–50.

Schrefler, B. A., Khoury, G. A., Gawin, D. and Majorana C. E., 2002. Thermo-hydro-mechanical modelling of high performance concrete at high temperatures, *Engineering Computations*, **19**, 787–819.

Schubert, W., Khanal, M. and Tomas, J., 2005. Impact crushing of particle-particle compounds: experiment and simulation, *International Journal of Mineral Processing*, **75**, 41–52.

Scott, D. R., 1996. Seismicity and stress rotation in a granular model of the brittle crust, *Nature*, **381**, 592–595.

Sethian, J. A., 1999. *Level Set Methods and Fast Marching Methods*, Cambridge University, Cambridge.

Sheldon, H. A. and Ord, A., 2005. Evolution of porosity, permeability and fluid pressure in dilatant faults post-failure: Implications for fluid flow and mineralization, *Geofluids*, **5**, 272–288.

Sih, G. C. and Macdonald, B., 1974. Fracture mechanics applied to engineering problems: strain energy density fracture criterion, *Engineering Fracture Mechanics*, **6**, 361–386.

Smooke M. D., Mcenally, C. S., Pfefferle, L. D., Hall, R. J. and Colket, M. B., 1999. Computational and experimental study of soot formation in a coflow, laminar diffusion flame, *Combustion and Flame*, **117**, 117–139.

Sorjonen-Ward, P., Zhang, Y. and Zhao, C., 2002. Numerical modelling of orogenic processes and mineralization in the south eastern part of the Yilgarn Craton, Western Australia, *Australian Journal of Earth Sciences*, **49**, 935–964.

Steefel, C. I. and Lasaga, A. C., 1990. Evolution of dissolution patterns: Permeability change due to coupled flow and reaction. In: Melchior, D. C. and Basset, R. L. (Eds.), *Chemical Modeling in Aqueous Systems II*, American Chemistry Society Symposium Series, **416**, 213–225.

References

Steefel, C. I. and Lasaga, A. C., 1994. A coupled model for transport of multiple chemical species and kinetic precipitation/dissolution reactions with application to reactive flow in single phase hydrothermal systems. *American Journal of Science*, **294**, 529–592.

Strayer, L. M. and Huddleston, P. J., 1997. Numerical modeling of fold initiation at thrust ramps, *Journal of Structural Geology*, **19**, 551–566.

Strayer, L. M. and Suppe, J., 2002. Out-of-plane motion of a thrust sheet during along-strike propagation of a thrust ramp: a distinct-element approach, *Journal of Structural Geology*, **24**, 637–650.

Suiker, A. S. J. and Fleck, N. A., 2004. Frictional collapse of granular assemblies, *Journal of Applied Mechanics-Transactions of the ASME*, **71**, 350–358.

Thornton, C., Ciomocos, M. T. and Adams, M. J., 1999. Numerical simulations of agglomerate impact breakage, *Powder Technology*, **105**, 74–82.

Tomas, J., Schreier, M., Groger, T. and Ehlers, S., 1999. Impact crushing of concrete for liberation and recycling, *Powder Technology*, **105**, 39–51.

Tornberg, A. K. and Engquist, B., 2003a. The segment projection method for interface tracking, *Communications on Pure and Applied Mathematics*, **56**, 47–79.

Tornberg, A. K. and Engquist, B., 2003b. Regularization techniques for numerical approximation of PDEs with singularities, *Journal of Scientific Computing*, **19**, 527–552.

Trevisan, O. V. and Bejan, A., 1987. Mass and heat transfer by high Rayleigh number convection in a porous medium heated from below, *International Journal of Mass Heat Transfer*, **30**, 2341–2356.

Turcotte, D. L. and Schubert, G., 1982. *Geodynamics: Applications of Continuum Physics to Geological Problems*, John Wiley & Sons, New York.

Walden, J., 1999. On the approximation of singular source terms in differential equations, *Numerical Methods for Partial Differential Equations*, **15**, 503–520.

Weinberg, R. F., 1996. Ascent mechanism of felsic magmas: News and views, *Transactions of the Royal Society of Edinburgh: Earth Sciences*, **87**, 95–103.

Wilde, A. R. and Wall, V. J., 1987. Geology of the Nabarlek Uranium deposit, Northern-territory, Australia, *Economic Geology*, **82**, 1152–1168.

Xu, T. F., Samper, J. Ayora, C., Manzano, M. and Custodio, E., 1999. Modelling of non-isothermal multi-component reactive transport in field scale porous media flow systems, *Journal of Hydrology*, **214**, 144–164.

Xu, T. F., Apps, J. A. and Pruess, K., 2004. Numerical simulation of CO2 disposal by mineral trapping in deep aquifers, *Applied Geochemistry*, **19**, 917–936.

Yang, J. W., 2006. Full 3-D numerical simulation of hydrothermal fluid flow in faulted sedimentary basins: Example of the McArthur Basin, Northern Australia. *Journal of Geochemical Exploration*, **89**, 440–444.

Yardley, B. W. D. and Bottrell, S. H., 1992. Silica mobility and fluid movement during metamorphism of the Connemara schists, Ireland, *Journal of Metamorphic Geology*, **10**, 453–464.

Yardley, B. W. D. and Lloyd, G. E., 1995. Why metasomatic fronts are really sides, *Geology*, **23**, 53–56.

Yeh, G. T. and Tripathi, V. S., 1989. A critical evaluation of recent developments in hydrogeochemical transport models of reactive multichemical components, *Water Resources Research*, **25**, 93–108.

Yeh, G. T. and Tripathi, V. S., 1991. A model for simulating transport of reactive multispecies components: Model development and demonstration, *Water Resources Research*, **27**, 3075–3094.

Zhang, Y., Hobbs, B. E., Ord, A. Barnicoat, A., Zhao, C. and Lin, G., 2003. The influence of faulting on host-rock permeability, fluid flow and ore genesis of gold deposits: a theoretical 2D numerical model, *Journal of Geochemical Exploration*, **78–79**, 279–284.

Zhang, Y., Lin, G., Wang, Y. J., Roberts, P. A. and Ord, A., 2007. Numerical modelling of deformation and fluid flow in the Shui-Kou-Shan Mineralisation District, Hunan Province, China, *Ore Geology Review*, **31**, 261–278.

Zhao, C. and Valliappan S., 1993a. Transient infinite elements for seepage problems in infinite media, *International Journal for Numerical and Analytical Methods in Geomechanics*, **17**, 324–341.

Zhao, C. and Valliappan S., 1993b. Mapped transient infinite elements for heat transfer problems in infinite media, *Computer Methods in Applied Mechanics and Engineering*, **108**, 119–131.

Zhao, C. and Valliappan S., 1994a. Transient infinite elements for contaminant transport problems in infinite media, *International Journal for Numerical Methods in Engineering*, **37**, 1143–1158.

Zhao, C. and Valliappan S., 1994b. Numerical modelling of transient contaminant migration problems in infinite porous fractured media using finite/infinite element technique: Theory and Parametric study, *International Journal for Numerical and Analytical Methods in Geomechanics*, **18**, 523–564.

Zhao, C., Xu, T. P. and Valliappan, S., 1994c. Numerical modelling of mass transport problems in porous media: A review, *Computers and Structures*, **53**, 849–860.

Zhao, C. and Steven, G. P., 1996a. Asymptotic solutions for predicted natural frequencies of two-dimensional elastic solid vibration problems in finite element analysis, *International Journal for Numerical Methods in Engineering*, **39**, 2821–2835.

Zhao, C. and Steven, G. P., 1996b. An asymptotic formula for correcting finite element predicted natural frequencies of membrane vibration problems, *Communications in Numerical Methods in Engineering*, **11**, 63–73.

Zhao, C. and Steven, G. P., 1996c. A practical error estimator for predicted natural frequencies of two-dimensional elastodynamic problems, *Engineering Computations*, **13**, 19–37.

Zhao, C., Steven, G. P. and Xie, Y. M., 1996d. Evolutionary natural frequency optimization of thin plate bending vibration problems, *Structural Optimization*, **11**, 244–251.

Zhao, C., Steven, G. P. and Xie, Y. M., 1996e. General evolutionary path for fundamental natural frequencies of structural vibration problems: towards optimum from below, *Structural Engineering and Mechanics*, **4**, 513–527.

Zhao C., Mühlhaus, H. B. and Hobbs, B. E., 1997a. Finite element analysis of steady-state natural convection problems in fluid-saturated porous media heated from below, *International Journal for Numerical and Analytical Methods in Geomechanics*, **21**, 863–881.

Zhao, C., Steven, G. P. and Xie, Y. M., 1997b. Effect of initial nondesign domain on optimal topologies of structures during natural frequency optimization, *Computers and Structures*, **62**, 119–131.

Zhao, C., Steven, G. P. and Xie, Y. M., 1997c. Evolutionary natural frequency optimization of 2D structures with additional nonstructural lumped masses, *Engineering Computations*, **14**, 233–251.

Zhao, C., Steven, G. P. and Xie, Y. M., 1997d. Evolutionary optimization of maximizing the difference between two natural frequencies of a vibrating structure, *Structural Optimization*, **13**, 148–154.

Zhao C., Hobbs, B. E. and Mühlhaus, H. B., 1998a. Finite element modelling of temperature gradient driven rock alteration and mineralization in porous rock masses, *Computer Methods in Applied Mechanics and Engineering*, **165**, 175–187.

Zhao C., Muhlhaus, H. B. and Hobbs, B. E., 1998b. Effects of geological inhomogeneity on high Rayleigh number steady-state heat and mass transfer in fluid-saturated porous media heated from below, *Numerical Heat Transfer*, **33**, 415–431.

Zhao, C., Steven, G. P. and Xie, Y. M., 1998c. A generalized evolutionary method for natural frequency optimization of membrane vibration problems in finite element analysis, *Computers and Structures*, **66**, 353–364.

Zhao, C., Hornby, P., Steven, G. P. and Xie, Y. M., 1998d. A generalized evolutionary method for numerical topology optimization of structures under static loading conditions, *Structural Optimization*, **15**, 251–260.

Zhao C., Hobbs, B. E. and Mühlhaus, H. B., 1998e. Analysis of pore-fluid pressure gradient and effective vertical-stress gradient distribution in layered hydrodynamic systems, *Geophysical Journal International*, **134**, 519–526.

Zhao, C., Hobbs, B. E. and Mühlhaus, H. B., 1999a. Finite element modelling of reactive mass transport problems in fluid-saturated porous media, *Communications in Numerical Methods in Engineering*, **15**, 501–513.

Zhao C., Hobbs, B. E. and Mühlhaus, H. B., 1999b. Theoretical and numerical analyses of convective instability in porous media with upward throughflow, *International Journal for Numerical and Analytical Methods in Geomechanics*, **23**, 629–646.

Zhao, C., Hobbs, B. E. and Mühlhaus, H. B., 1999c. Effects of medium thermoelasticity on high Rayleigh number steady-state heat transfer and mineralization in deformable fluid-saturated porous media heated from below, *Computer Methods in Applied Mechanics and Engineering*, **173**, 41–54.

Zhao C., Hobbs, B. E., Mühlhaus, H. B. and Ord, A., 1999d. Finite element analysis of flow patterns near geological lenses in hydrodynamic and hydrothermal systems, *Geophysical Journal International*, **138**, 146–158.

Zhao, C., Hobbs, B. E., Mühlhaus, H. B. and Ord, A., 1999f. A consistent point-searching algorithm for solution interpolation in unstructured meshes consisting of 4-node bilinear quadrilateral elements, *International Journal for Numerical Methods in Engineering*, **45**, 1509–1526.

Zhao, C., Hobbs B. E., Baxter, K., Mühlhaus H. B. and Ord, A., 1999g. A numerical study of pore-fluid, thermal and mass flow in fluid-saturated porous rock basins, *Engineering Computations*, **16**, 202–214.

Zhao, C., Hobbs, B. E. and Mühlhaus, H. B., 2000a. Finite element analysis of heat transfer and mineralization in layered hydrothermal systems with upward throughflow, *Computer Methods in Applied Mechanics and Engineering*, **186**, 49–64.

Zhao, C., Hobbs, B. E., Mühlhaus, H. B., Ord, A. and Lin, G., 2000b. Numerical modelling of double diffusion driven reactive flow transport in deformable fluid-saturated porous media with particular consideration of temperature-dependent chemical reaction rates, *Engineering Computations*, **17**, 367–385.

Zhao, C., Hobbs, B. E., Mühlhaus, H. B. and Ord, A., 2000c. Finite element modelling of dissipative structures for nonequilibrium chemical reactions in fluid-saturated porous media, *Computer Methods in Applied Mechanics and Engineering*, **184**, 1–14.

Zhao, C., Hobbs, B. E., Mühlhaus, H. B., Ord, A. and Lin, G., 2001a. Finite element modelling of three-dimensional convection problems in pore-fluid saturated porous media heated from below, *Communications in Numerical Methods in Engineering*, **17**, 101–114.

Zhao C., Lin, G., Hobbs, B. E., Mühlhaus, H. B., Ord, A. and Wang, Y., 2001b. Finite element modelling of heat transfer through permeable cracks in hydrothermal systems with upward throughflow, *Engineering Computations*, **18**, 996–1011.

Zhao, C., Hobbs, B. E., Walshe, J. L., Mühlhaus, H. B. and Ord, A., 2001c. Finite element modeling of fluid-rock interaction problems in pore-fluid saturated hydrothermal/sedimentary basins, *Computer Methods in Applied Mechanics and Engineering*, **190**, 2277–2293.

Zhao, C., Hobbs, B. E., Mühlhaus, H. B. and Ord, A., 2001d., Finite element modelling of rock alteration and metamorphic process in hydrothermal systems, *Communications in Numerical Methods in Engineering*, **17**, 833–843.

Zhao, C., Lin, G., Hobbs, B. E., Wang, Y., Mühlhaus, H. B. and Ord, A., 2002a. Finite element modelling of reactive fluids mixing and mineralization in pore-fluid saturated hydrothermal/sedimentary basins, *Engineering Computations*, **19**, 364–387.

Zhao, C., Hobbs, B. E., Mühlhaus, H. B., Ord, A. and Lin, G., 2002b. Computer simulations of coupled problems in geological and geochemical systems, *Computer Methods in Applied Mechanics and Engineering*, **191**, 3137–3152.

Zhao C., Hobbs, B. E., Mühlhaus, H. B., Ord, A. and Lin, G., 2002c. Analysis of steady-state heat transfer through mid-crustal vertical cracks with upward throughflow in hydrothermal systems, *International Journal for Numerical and Analytical Methods in Geomechanics*, **26**, 1477–1491.

Zhao, C., Hobbs, B. E., Mühlhaus, H. B., Ord, A. and Lin, G., 2003a. Finite element modeling of three-dimensional steady-state convection and lead/zinc mineralization in fluid-saturated rocks, *Journal of Computational Methods in Science and Engineering*, **3**, 73–89.

Zhao, C., Hobbs, B. E., Ord, A., Lin, G. and Mühlhaus, H. B., 2003b. An equivalent algorithm for simulating thermal effects of magma intrusion problems in porous rocks, *Computer Methods in Applied Mechanics and Engineering*, **192**, 3397–3408.

Zhao, C., Hobbs, B. E., Ord, A., Mühlhaus, H. B. and Lin, G., 2003c. Effect of material anisotropy on the onset of convective flow in three-dimensional fluid-saturated faults, *Mathematical Geology*, **35**, 141–154.

Zhao, C., Hobbs, B. E., Mühlhaus, H. B., Ord, A. and Lin, G., 2003d. Convective instability of three-dimensional fluid-saturated geological fault zones heated from below, *Geophysical Journal International*, **155**, 213–220.

Zhao, C., Lin, G., Hobbs, B. E., Ord, A., Wang, Y. and Mühlhaus, H. B., 2003e. Effects of hot intrusions on pore-fluid flow and heat transfer in fluid-saturated rocks, *Computer Methods in Applied Mechanics and Engineering*, **192**, 2007–2030.

Zhao, C., Hobbs, B. E., Ord, A., Peng, S., Mühlhaus, H. B. and Liu, L., 2004. Theoretical investigation of convective instability in inclined and fluid-saturated three-dimensional fault zones, *Tectonophysics*, **387**, 47–64.

Zhao, C., Hobbs, B. E., Ord, A., Peng, S., Mühlhaus, H. B. and Liu, L., 2005a. Numerical modeling of chemical effects of magma solidification problems in porous rocks, *International Journal for Numerical Methods in Engineering*, **64**, 709–728.

Zhao, C., Hobbs, B. E., Ord, A., Peng, S., Mühlhaus, H. B. and Liu, L., 2005b. Double diffusion-driven convective instability of three-dimensional fluid-saturated geological fault zones heated from below, *Mathematical Geology*, **37**, 373–391.

Zhao, C., Hobbs, B. E., Ord, A., Kuhn, M., Mühlhaus, H. B. and Peng, S., 2006a. Numerical simulation of double-diffusion driven convective flow and rock alteration in three-dimensional fluid-saturated geological fault zones, *Computer Methods in Applied Mechanics and Engineering*, **195**, 2816–2840.

Zhao, C., Hobbs, B. E., Ord, A. and Hornby, P., 2006b. Chemical reaction patterns due to fluids mixing and focusing around faults in fluid-saturated porous rocks, *Journal of Geochemical Exploration*, **89**, 470–473.

Zhao, C., Hobbs, B. E., Hornby, P., Ord, A. and Peng, S., 2006c. Numerical modelling of fluids mixing, heat transfer and non-equilibrium redox chemical reactions in fluid-saturated porous rocks, *International Journal for Numerical Methods in Engineering*, **66**, 1061–1078.

Zhao, C., Hobbs, B. E., Ord, A., Hornby, P., Peng, S. and Liu, L., 2006d. Theoretical and numerical analyses of pore-fluid flow patterns around and within inclined large cracks and faults, *Geophysical Journal International*, **166**, 970–988.

Zhao, C., Hobbs, B. E., Ord, A., Peng, S., Liu, L. and Mühlhaus, H. B., 2006e. Analytical solutions for pore-fluid flow focusing within inclined elliptical inclusions in pore-fluid-saturated porous rocks: Solutions derived in an elliptical coordinate system, *Mathematical Geology*, **38**, 987–1010.

Zhao, C., Nishiyama, T. and Murakami, A., 2006 f. Numerical modelling of spontaneous crack generation in brittle materials using the particle simulation method, *Engineering Computations*, **23**, 566–584.

Zhao, C., Hobbs, B. E., Ord, A., Hornby, P., Peng, S. and Liu, L., 2007a. Mineral precipitation associated with vertical fault zones: The interaction of solute advection, diffusion and chemical kinetics, *Geofluids*, **7**, 3–18.

Zhao, C., Hobbs, B. E., Ord, A., Hornby, P., Peng, S. and Liu, L., 2007b. Particle simulation of spontaneous crack generation problems in large-scale quasi-static systems, *International Journal for Numerical Methods in Engineering*, **69**, 2302–2329.

Zhao, C., Hobbs, B. E., Ord, A., Peng, S. and Liu, L., 2007c. An upscale theory of particle simulation for two-dimensional quasi-static problems, *International Journal for Numerical Methods in Engineering*, **72**, 397–421.

References

Zhao, C., Hobbs, B. E., Ord, A., Robert, P. A., Hornby, P. and Peng, S., 2007d. Phenomenological modeling of crack generation in brittle crustal rocks using the particle simulation method, *Journal of Structural Geology*, **29**, 1034–1048.

Zhao, C., Hobbs, B. E. and Ord, A., 2008a. Investigating dynamic mechanisms of geological phenomena using methodology of computational geosciences: an example of equal-distant mineralization in a fault, *Science in China D-Series: Earth Sciences*, **51**, 947–954.

Zhao, C., Hobbs, B. E., Ord, A., Peng, S. and Liu, L., 2008b. Inversely-mapped analytical solutions for flow patterns around and within inclined elliptic inclusions in fluid-saturated rocks, *Mathematical Geosciences*, **40**, 179–197.

Zhao, C., Hobbs, B. E., Ord, A., Hornby, P., Mühlhaus, H. B. and Peng, S., 2008c. Theoretical and numerical analyses of pore-fluid-flow focused heat transfer around geological faults and large cracks, *Computers and Geotechnics*, **35**, 357–371.

Zhao, C., Hobbs, B. E., Ord, A., Hornby, P. and Peng, S., 2008d. Effect of reactive surface areas associated with different particle shapes on chemical-dissolution front instability in fluid-saturated porous rocks, *Transport in Porous Media*, **73**, 75–94.

Zhao, C., Hobbs, B. E., Ord, A., Hornby, P. and Peng, S., 2008e. Morphological evolution of three-dimensional chemical dissolution front in fluid-saturated porous media: A numerical simulation approach, *Geofluids*, **8**, 113–127.

Zhao, C., Hobbs, B. E., Hornby, P., Ord, A., Peng, S. and Liu, L., 2008 f. Theoretical and numerical analyses of chemical-dissolution front instability in fluid-saturated porous rocks, *International Journal for Numerical and Analytical Methods in Geomechanics*, **32**, 1107–1130.

Zhao, C., Hobbs, B. E., Ord, A., Peng, S., 2008 g. Particle simulation of spontaneous crack generation associated with the laccolithic type of magma intrusion processes, *International Journal for Numerical Methods in Engineering*, **75**, 1172–1193.

Zhao, P. H. and Heinrich, J. C., 2002. Approximation to the interface velocity in phase change front tracking, *Communications in Numerical Methods in Engineering*, **18**, 77–88.

Zienkiewicz, O. C., 1977. *The Finite Element Method*, McGraw-Hill, London.

Zienkiewicz, O. C. and Zhu, J. Z., 1991. Adaptivity and mesh generation, *Int. J. Num. Meth. Eng.*, **32**, 783–810.

Zimmerman, R. W., 1996. Effective conductivity of a two-dimensional medium containing elliptical inhomogeneities, *Proceedings of the Royal Society of London A*, **452**, 1713–1727.

Index

A

Advection, 74, 77, 80, 82, 121–123, 134, 136–142, 148, 150
Algorithm, 4–6, 37–38, 42–43, 45, 51, 56, 59, 71, 73, 76–77, 81–82, 88, 95, 109, 111, 115, 128, 153–156, 158–160, 162–164, 166–168, 176, 192–193, 221–225
Application, 1–2, 22, 43, 56, 59, 61, 65, 70, 82, 93, 115, 134, 155, 163, 168–169, 178, 192, 194, 199, 210, 219, 225
Approach, 6, 8–9, 13–16, 19–23, 28–29, 32, 37, 42, 59, 100, 102–103, 107, 122, 134, 136, 156, 177–178, 185, 222
Approximate, 4, 14, 28, 59, 78, 84, 96, 109, 110–112, 115, 128, 146, 148, 165, 169, 199, 204, 216–217
Assumption, 9, 40, 65, 84, 161, 195
Asymptotic approach, 6–9, 13–16, 19–23, 28–29, 32, 42, 100, 167, 222

B

Benchmark, 4–5, 16–19, 37, 96, 109, 111–112, 115, 164–165, 168, 222–223, 225
Bifurcation, 8, 20, 222
Boundary condition, 16, 18, 20, 23, 30, 60, 101, 103, 106, 111–113, 129, 136, 158, 168–169, 175, 190, 205, 216
Buoyancy, 7, 69

C

Concentration, 39–40, 42, 60–61, 63, 65–70, 74, 77–80, 84, 88, 93, 95–97, 99–100, 102, 104, 108–112, 114–117, 124–125, 128–129, 132–133, 136–137, 142, 146–151, 156–157, 163, 169, 171–173, 223
Conduction, 1, 7
Conductivity, 9–10, 30, 51, 61, 129, 157, 165
Conservation, 3, 78–79
Continuity, 104, 157
Convection, 7–9, 14–16, 18, 22, 25, 38, 51, 60, 77, 80, 134
Convergence, 5, 110–111, 156
Coordinate, 10–11, 28–29, 38, 43–47, 50, 100–101, 105, 157–158, 179–181, 222
Crack, 6, 122–123, 153–154, 175–179, 184, 192–195, 198, 200–201, 213, 216–220, 225
Crust, 2, 6–7, 29, 34, 38, 65, 73, 118, 121, 123, 128, 130–131, 136–137, 153–155, 168–169, 172, 175, 177, 204, 210–211, 216–218, 220–224

D

Darcy's law, 7, 9, 95–96, 123
Darcy velocity, 74, 78, 80, 83–84, 97, 102, 138
Decoupling, 6, 121, 123–126, 128, 133–134, 224
Density, 7, 9, 30, 39–40, 51, 61, 69, 97–100, 102, 112, 114, 118, 123–124, 129, 157, 160, 166–167, 170, 176, 197, 199, 204, 212, 216
Diffusivity, 61, 74, 97, 112, 129, 156–157, 162
Dispersion, 74, 77–78, 82, 84, 108, 115, 121, 123, 129, 136–140, 142–145, 148, 150
Dispersivity, 69, 78
Dissolution, 6, 73–77, 79–80, 84, 88, 93, 95–105, 107–109, 111–113, 115–119, 121, 223–224
Double diffusion, 40, 73

E

Energy, 3, 7–8, 13–14, 40, 61, 176
Equation, 4, 7–10, 13–14, 16–18, 28, 39–41, 43, 45, 47–50, 69–70, 74–75, 77–83, 88, 97, 99, 101, 103–104, 106–110, 122–124, 126–128, 133, 138, 144–145, 148, 156–157, 159, 162, 180, 183–184, 186–187, 192, 195–198, 223–224

E

Equilibrium, 6, 40–41, 61, 74–75, 79, 83–84, 95–96, 98–100, 102, 108, 112, 115, 121–122, 124, 126–128, 133–134, 136–140, 142–146, 148, 150, 157, 160, 187–188, 191–193, 216, 224
Expansion coefficient, 9, 30, 39, 41, 51, 56, 61, 124, 129

F

Fault, 2, 6, 122–123, 128–134, 136–137, 140–146, 148, 150, 195, 216–220, 224
Faulting process, 216
Fick's law, 96, 123
Finite difference, 4–5, 38, 42, 96, 109–110, 154, 178, 183–184, 218, 223
Finite element, 11, 13, 19, 22, 24, 30, 160, 164, 168
Flow, 22, 28, 37, 56, 59, 74, 112, 179, 189
Fluids mixing, 123
Focusing, 96, 118, 129–134, 136–138, 140–142, 144–146, 148, 150, 224
Folding process, 210, 213–215
Fourier's law, 123

G

Gravity, 7, 9, 14–15, 21, 29–30, 33, 39–40, 124, 197–198, 213, 217, 222
Green-Gauss theorem, 12

H

Heat, 6–9, 12–13, 22, 29–30, 32, 37–40, 51, 56, 59–61, 73, 121–124, 128–129, 137, 145, 153–160, 162–165, 168, 222, 224–225
Heat transfer, 6, 38–39, 51, 56, 59, 73, 121–124, 128, 137, 145, 154–157, 164, 168
Hydrothermal system, 2, 6, 33, 38, 40, 59–65, 69–70, 73, 137, 221–222

I

Initial condition, 82–84, 128, 169
Instability, 6, 75, 95–96, 100, 105, 107–108, 115–116, 223–224
Intrusion, 153–155, 166, 168–173
Inverse problem, 185

K

Kilometer, 4, 156, 177, 186, 205–206, 208, 211–214
Knowledge, 73, 88, 154, 178, 225

L

Length scale, 153–156
Linear problem, 8, 97

M

Magma intrusion, 153–155, 165, 168–173, 224
Mineral, 2, 6–8, 25, 34, 38, 59, 65, 73–80, 84, 95–99, 108, 112, 118, 121–122, 124, 134, 136–138, 141–146, 150, 153–156, 172–173, 175, 177, 221, 223–224
Mineralization, 6–8, 25, 34, 38, 59, 65, 73, 96, 118, 121–122, 124, 136, 153–154, 156, 173, 175, 177, 221

N

Non-equilibrium, 6, 40, 121, 127, 148, 224
Numerical method, 1, 4–5, 8, 30, 59, 73, 76, 96, 109, 111, 124, 127–128, 153–156, 175–176, 218–219, 221–222, 224
Numerical solution, 1, 4–5, 20–23, 31, 33, 43, 50, 59, 61, 65, 70, 82, 88, 93, 96, 102, 104, 109, 112–115, 130–133, 165–167, 188, 222, 224

O

Oberbeck-Boussinesq approximation, 7, 9, 123
Ore body formation, 2, 34, 59, 65, 73, 96, 121–122, 124, 153–154, 156, 172–173, 175, 221–222
Ore deposits, 73, 96, 118, 121

P

Particle simulation, 5–6, 175–179, 183–184, 188–189, 193–194, 196–201, 204–205, 208, 210, 213, 215–217, 219, 225
Peclet number, 76–77, 81–82, 223
Permeability, 10, 23, 30, 41–42, 51, 60–61, 64–65, 75, 95–99, 105, 108, 112, 126, 129, 131, 135, 144, 146
Pore-fluid, 22, 28, 37, 56, 59
Porosity, 9, 30, 39–42, 51, 56, 60–61, 64–65, 74–77, 79–82, 84, 92–93, 95–100, 103–105, 109–113, 115–116, 118, 124–126, 138, 144, 199, 204, 223
Porous medium, 7–10, 14–15, 17–19, 23, 25, 31–36, 38–41, 56–61, 65, 69, 74, 78–79, 81, 84, 88, 95–100, 108, 111–113, 115–119, 124, 138, 222, 224
Precipitation, 6, 73, 76, 80, 84, 88, 93, 97–99, 121, 134, 136–138, 140–146, 148, 150
Pressure gradient, 73–74, 100, 111, 115, 137, 144, 146, 148, 150
Procedure, 1, 6–8, 13–14, 16–17, 19–23, 28, 32, 42, 51, 96, 109–111, 113, 115–116, 118, 121–126, 128, 130, 133–134, 184, 189, 191, 198, 201–205, 217, 221–225
Progressive asymptotis approach, 6–8, 13–14, 16, 19–20, 22, 28–29, 32, 42, 222

Index

R

Rayleigh number, 7, 10, 15, 18–21, 23, 25, 29, 38, 42, 51, 61, 222

Response, 177–178, 186, 188–189, 198–199, 201, 203–205, 207–208, 214–215, 218

Rock, 2, 6, 9, 29, 73–93, 95–97, 108, 121–126, 128–129, 131, 133–134, 136–137, 140, 144, 146, 148, 150, 153, 155, 157–159, 161–163, 165–169, 172–173, 175–177, 179, 186, 200, 204–205, 208, 217, 220, 223–225

S

Shape function, 13, 46

Solution, 1, 4–6, 8, 14–23, 28, 30–33, 37–38, 42–45, 47–56, 59–61, 65, 70, 73, 79, 82, 88, 93, 95–96, 99–100, 102, 104–106, 108–115, 123, 127–128, 130–134, 138, 141, 143, 156–157, 161–167, 183–184, 187–188, 192–193, 198, 201, 203, 217, 222–224

Source, 1, 6, 37, 82, 96–99, 124–128, 133, 135, 153, 155–160, 162–165, 168

Steady-state, 6–9, 14, 16, 20, 38–39, 101, 222

Streamline, 23, 26–28, 117–119, 129–130, 146

T

Temperature, 7, 9–14, 17, 20, 22, 30–33, 39–42, 51–52, 54, 56–62, 65, 69, 121, 123–124, 129, 136–137, 155–158, 160–163, 165, 168–172, 175

Term splitting, 6, 73, 77, 81–82, 223

Thermal effect, 41, 61, 65, 154, 156, 166

Transport, 6, 8, 25, 37–40, 59, 61, 69, 73, 75–80, 82–83, 88, 95–97, 99, 105, 107–108, 111, 118, 121–124, 126–128, 133–134, 137–138, 148, 154–157, 164, 222–224

U

Upper crust, 2, 6, 29, 34, 38, 65, 73, 118, 121, 123, 154–155, 168–169, 172, 175, 177, 210, 221–224

Upscale theory, 178, 194, 196, 198–199, 204–205, 208, 211–215, 225

Upward throughflow, 128

V

Validation, 3, 221

Velocity, 8–15, 20, 23, 25, 28–29, 31–33, 39, 42, 60, 62, 69, 74, 76, 78–80, 82–84, 97, 99, 102, 124, 129, 131–132, 138–139, 143–144, 146, 148, 150, 181–182, 189, 191, 193, 201, 205, 222

Verification, 4, 19, 23, 51, 56, 111, 128, 146, 163, 168

Vertical stress, 56

Viscosity, 9, 30, 39, 51, 61, 97, 124, 129, 144

Z

Zhao number, 6, 73–75, 107–108, 111–112, 115, 223–224

Zone, 118, 129, 133–134, 136–137, 140–146, 148, 150